£ 16.95

Safety and health in ports

GW00655898

ILO code of practice

Safety and health in ports

International Labour Office   Geneva

Copyright © International Labour Organization 2005
First published 2005

Publications of the International Labour Office enjoy copyright under Protocol 2 of the Universal Copyright Convention. Nevertheless, short excerpts from them may be reproduced without authorization, on condition that the source is indicated. For rights of reproduction or translation, application should be made to ILO Publications (Rights and Permissions), International Labour Office, CH-1211 Geneva 22, Switzerland. The International Labour Office welcomes such applications.

Libraries, institutions and other users registered in the United Kingdom with the Copyright Licensing Agency, 90 Tottenham Court Road, London W1T 4LP [Fax: (+44) (0)20 7631 5500; email: cla@cla.co.uk], in the United States with the Copyright Clearance Center, 222 Rosewood Drive, Danvers, MA 01923 [Fax: (+1) (978) 750 4470; email: info@copyright.com] or in other countries with associated Reproduction Rights Organizations, may make photocopies in accordance with the licences issued to them for this purpose.

ILO

Safety and health in ports. ILO code of practice
Geneva, International Labour Office, 2005

Occupational safety, occupational health, port, docker, code of practice. 13.04.2
ISBN 92-2-115287-1

Also published in French: *Sécurité et santé dans les ports. Recueil de directives pratiques du BIT* (ISBN 92-2-215287-5), Geneva, 2005); and in Spanish: *Seguridad y salud en los puertos. Repertorio de recomendaciones prácticas de la OIT* (ISBN 92-2-315287-9, Geneva, 2005).

ILO Cataloguing in Publication Data

The designations employed in ILO publications, which are in conformity with United Nations practice, and the presentation of material therein do not imply the expression of any opinion whatsoever on the part of the International Labour Office concerning the legal status of any country, area or territory or of its authorities, or concerning the delimitation of its frontiers.

The responsibility for opinions expressed in signed articles, studies and other contributions rests solely with their authors, and publication does not constitute an endorsement by the International Labour Office of the opinions expressed in them.

Reference to names of firms and commercial products and processes does not imply their endorsement by the International Labour Office, and any failure to mention a particular firm, commercial product or process is not a sign of disapproval.

ILO publications can be obtained through major booksellers or ILO local offices in many countries, or direct from ILO Publications, International Labour Office, CH-1211 Geneva 22, Switzerland. Catalogues or lists of new publications are available free of charge from the above address, or by email: pubvente@ilo.org
Visit our website: www.ilo.org/publns

Photocomposed by the ILO, Geneva, Switzerland
Printed in Malta

DTP
INT

Preface

This ILO code of practice on safety and health in ports replaces two former ILO publications: *Guide to safety and health in dock work* (1976) and *Safety and health in dock work*, An ILO code of practice (second edition, 1977). The text was adopted by a Meeting of Experts held in Geneva from 8 to 17 December 2003. The Governing Body of the ILO, at its 287th Session in June 2003, approved the composition of the meeting of experts. In accordance with this decision, the meeting was attended by 12 experts nominated by Governments (Australia, Brazil, Canada, China, Egypt, Germany, Nigeria, Panama, Philippines, Spain, United Kingdom, United States), 12 experts nominated by the Employers and 12 experts nominated by the Workers. Expert observers from other governments, and observers from a number of intergovernmental and non-governmental organizations also attended.

This code of practice is not a legally binding instrument. It is not intended to replace national laws and regulations or to affect the fundamental principles and rights of workers provided by ILO instruments.

The practical recommendations in this code are intended to provide relevant guidance to ILO constituents and all those responsible for or involved in the management, operation, maintenance and development of ports.

It is hoped that this code will help to raise the profile of safety and health issues in ports in all parts of the world, and will encourage more countries to ratify the Occupational Safety and Health (Dock Work) Convention, 1979 (No. 152), or otherwise implement its provisions.

The ILO wishes to record its grateful thanks to Circlechief AP and the Circlechief team of John Alexander and Mike Compton of the United Kingdom for their comprehensive work in developing the code of practice and assisting the meeting of experts in its discussions. The ILO also gratefully acknowledges the financial support provided by Through Transport Mutual Services (UK) Ltd. (TT CLUB) for the preparation of the technical illustrations included in this code.

List of participants

Experts nominated by Governments:

Mr. John Kilner, Assistant Secretary, Maritime Security, Department of Transport and Regional Services, Canberra (Australia).

Mr. John Platts, Special Advisor, Marine Security, Transport Canada, Security and Emergency and Preparedness Directorate, Ottawa (Canada).

> *Adviser:*

> Ms. Lynn Young, Director, Human Resources Development Canada (HRDC), Ottawa.

Mr. Daltro D'Arisbo, Labour Office Auditor – FISCAL, Porto Alegre (Brazil).

> *Adviser:*

> Captain Darlei Pinheiro, Mission Officer, Brazilian Permanent Representation to the IMO, London.

Mr. Ye Hongjun, Division Chief, Department of Worker Transport Administration, Ministry of Communication, Beijing (China).

> *Advisers:*

> Mr. Xu Yi, Director, Department of Human Resources, Ministry of Communication, Beijing.

Ms. Zhao Xiaoliang, Official, Department of International Cooperation, Ministry of Communication, Beijing.

Mr. Tarek Hassan Ibrahim Sharef Eldin, Director, Industrial Health and Safety Institute, Cairo (Egypt).

Advisers:

Mr. Hazem Abdel Hazem Halim, Head of Maritime Sector Administration, Alexandria.

Ms. Nadia El-Gazzar, Labour Counsellor, Permanent Mission of Egypt, Geneva.

Mr. Achim Sieker, Junior Unit Leader, Federal Ministry of Economics and Labour, Bonn (Germany).

Adviser:

Ms. Ute Bödecker, Executive for Legal and Security Matters, Ministry of the Interior, Hamburg.

Mr. Mobolaji Olurotimi Banjo, Acting Director, Inspectorate Department, Federal Ministry of Labour and Productivity, Abuja (Nigeria).

Advisers:

Mr. Wali Mansoor Kurawa, Director of Maritime Services, Federal Ministry of Transport, Abuja.

Mr. Sotonye Inyeinengi-Etomi, Special Assistant to the Honourable Minister of Transport, Federal Ministry of Transport, Abuja.

Dr. O.C. Nathaniel, Deputy Director, Joint Maritime Labour Industrial Council, Lagos.

Ms. Ifeoma Christina Nwankwo, Deputy Director, Federal Ministry of Labour and Productivity, Abuja.

Mr. John Idakwoji, Assistant Chief Administrative Officer, Abuja.

Mr. Audu Igho, Administrative Officer, Joint Maritime Labour Industrial Council, Lagos.

Mr. Henry A. Ajetunmobi, Assistant General Manager Security, Nigerian Ports Authority, Lagos.

Ms. Julissa Tejada de Humphrey, Director, Institutional Office of Human Resources, Ministry of Labour and Manpower Development, Panama (Panama).

Mr. Gerardo S. Gatchalian, Supervising Labor and Employment Officer, Bureau of Working Conditions, Department of Labor and Employment, Manila (Philippines).

Advisers:

Mr. Benjamin B. Cecilio, Assistant General Manager for Operations, Philippine Ports Authority, Manila.

Ms. Yolanda C. Porschwitz, Labor Attaché, Permanent Mission of the Philippines, Geneva.

Mr. D. Juan Ramón Bres, Inspector, Labour and Social Security Inspectorate, Cadiz (Spain).

Adviser:

Mr. Pedro J. Roman Nuñez, Safety and Security Department Manager, Spanish Ports Administration, Madrid.

Mr. Graeme Henderson, Head of Marine and Civil Contingencies Section, Health and Safety Executive, London (United Kingdom).

Advisers:

Mr. David Carter, Deputy Head and Health and Safety Executive, Transport Safety Division, Marine, Aviation and Civil Contingencies Section, London.

Mr. Ashley Reeve, Head of Maritime Security Operations, Department of Transport, London.

Mr. Charles Thomas Pope, Area Director, US Department of Labor Occupational Safety and Health Administration (OSHA), Norfolk (United States).

Advisers:

Captain Jon S. Helmick, USMS, Director, Logistics and Intermodal Transportation Program, United States Merchant Marine Academy, Kings Point.

Captain David L. Scott, Chief, Office of Operating and Environmental Standards, Washington, DC.

Mr. John W. Chamberlin, First Secretary, Permanent Mission of the United States of America, Chambésy.

Experts nominated by the Employers:

Mr. Abdou Ba, Director of the Port Manpower Bureau, Unions of Cargo-Handling Enterprises of Senegal Ports, Mole (Sénégal).

Mr. Francis Bertrand, Director of Human Resources and Legal Affairs, International Organisation of Employers (IOE), Nantes (France).

Mr. Guido Marcelo Bocchio Carbajal, Legal Superintendent, Southern Peru Copper Corporation, Lima (Peru).

Mr. Joseph J. Cox, President, Chamber of Shipping of America, Washington, DC (United States).

Ms. Lynne Harrison, Human Resources Manager, Port of Napier Ltd., Napier (New Zealand).

Mr. Dierk Lindemann, Managing Director, German Shipowners' Association, Hamburg (Germany).

Mr. Claes Olmarker, Port Security Officer, Port of Göthenburg, Göthenburg (Sweden).

Mr. Usman Husein Punjwani, Seaboard Services Partner, Seaboard Services, Karachi (Pakistan).

Mr. Dahari Ujud, Senior Manager, Ancilliary Services, Port Klang (Malaysia).

Mr. Michael Joseph Van der Meer, General Manager, Port Authority Division, Namibian Ports Authority, Walvis Bay (Namibia).

Mr. Pieter M. Van der Sluis, Adviser Social Affairs, General Employers' Organization, Nieuwerherh (The Netherlands).

Adviser:

Mr. Fer M.J. Van de Laar, Chairman, International Association of Ports and Harbours, Environment and Marine Operations Committee, Amsterdam Port Authority, Amsterdam.

Mr. Alexander Zaitsev, President, Association of Ports and Shipowners of River Transport, Moscow (Russian Federation).

Experts nominated by the Workers:

Mr. Gary Brown, Port Security, AFL-CIO, Fife, Washington State (United States).

Mr. Marcel Carlstedt, Swedish Transport Workers' Union, Stockholm (Sweden).

Mr. P.M. Mohamed Haneef, Working President, Cochin Port Staff Association, Kochi (India).

Mr. Knud Hansen, Secretary, General Workers' Union, Copenhagen (Denmark).

Mr. Albert François Le Monnier, Third Vice-President, Safety Coordinator, International Longshore and Warehouse Union (ILWU), Vancouver (Canada).

Mr. Peter Lovkvist, Co-Sweden, Valbo (Sweden).

Mr. Kees Marges, Union Adviser, FNV Bondgeroter Netherlands, London (United Kingdom).

Ms. Veronica Mesatywa, National Sector Coordinator, Maritime Industry, South African Transport and Allied Workers' Union (SATAWU), Johannesburg (South Africa).

Mr. Leal Sundet, International Longshore and Warehouse Union (ILWU), California (United States).

Mr. James Trevor Tannock, Deputy National Secretary, Maritime Union of Australia, Sydney (Australia).

Mr. Kenji Yasuda, President, National Council of Dockworkers' Union of Japan (ZENKOKU-KOWAN), Tokyo (Japan).

Advisers:

Mr. Yuji Iijima, Chief of European Office, All Japan Seamen's Union, London.

Mr. Shimpei Suzuki, General Secretary, National Council of Dockworkers' Union of Japan, Tokyo.

In addition to the experts, Government observers from three member States, as well as ten representatives of intergovernmental and non-governmental organizations, attended.

Government observers:

Mr. Harri Halme, Senior Safety Officer, Ministry of Social Affairs and Health, Tampere (Finland).

Mr. Mario Alvino, Official, Ministry of Labour and Social Policy, Rome (Italy).

Ms. Asiye Türker, Senior Engineer, Head of Delegation, The Prime Ministry Undersecretariat for Maritime Affairs, Ankara (Turkey).

Representatives of intergovernmental organizations:

Mr. W. Elsner, Head of Port Unit, European Union (EU), Brussels.

Mr. Diego Teurelincx, Port Unit, EU, Brussels.

Mr. Christopher C. Trelawny, Senior Technical Officer, Navigational Safety and Maritime and Security Section, Maritime Safety Division, International Maritime Organization (IMO), London.

Mr. Viatcheslav Novikov, Economic Affairs Officer, United Nations Economic Commission for Europe (UNECE), Geneva.

Representatives of non-governmental international organizations:

Mr. Fer M.J. Van de Laar, Chairman, International Association of Ports and Harbours (IAPH), Environment and Marine Operations Committee, Amsterdam Port Authority, Amsterdam.

Mr. John Nicholls, Director, ICHCA International, Romford, Essex.

Mr. Brian Parkinson, Adviser, International Shipping Federation (ISF), London.

Mr. Dan Cunniah, Director, Geneva Office, International Confederation of Free Trade Unions (ICFTU), Geneva.

Ms. Anna Biondi, Assistant Director, Geneva Office, ICFTU, Geneva.

Mr. Jean Dejardin, Adviser, International Organisation of Employers (IOE), Geneva.

ILO representatives:

Ms. Cleopatra Doumbia-Henry, Director, Sectoral Activities Department.

Mr. Dani Appave, Maritime Specialist, Maritime Activities, Sectoral Activities Department.

Mr. Marios Meletiou, Technical Specialist (Ports and Transport), Maritime Activities, Sectoral Activities Department.

Contents

List of figures

List of abbreviations and acronyms

ACEP	approved continuous examination programme
APELL	Awareness and Preparedness for Emergencies at Local Level (UNEP/IMO)
BC Code	*Code of Practice for Solid Bulk Cargoes* (IMO)
BLU Code	*Code of Practice for the Safe Loading and Unloading of Bulk Carriers* (IMO)
CCTV systems	closed-circuit television surveillance systems
CSC	International Convention for Safe Containers, 1972 (IMO)
CTU	cargo transport unit
EmS	*Emergency Schedules* (supplement to the *IMDG Code*)
FIBC	flexible intermediate bulk container
IAPH	International Association of Ports and Harbours
ICS	International Chamber of Shipping
IEC	International Electrotechnical Commission
ILO	International Labour Organization
IMDG Code	*International Maritime Dangerous Goods Code* (IMO)
IMO	International Maritime Organization
ISGOTT	*International Safety Guide for Oil Tankers and Terminals* (ICS/IAPH)
ISO	International Organization for Standardization
ISPS Code	*International Ship and Port Facility Security Code* (IMO)
LED	light-emitting diode
LNG	liquefied natural gas
LPG	liquefied petroleum gas

MEWP	mobile elevating work platforms (or "cherry pickers")
MFAG	*Medical First Aid Guide* (supplement to the *IMDG Code*)
OECD	Organisation for Economic Co-operation and Development
OCIMF	Oil Companies International Marine Forum
OSH	occupational safety and health
PAH	polycyclic aromatic hydrocarbons
PDP	ILO Portworker Development Programme
PPE	personal protective equipment
PSN	Proper Shipping Name
RMG	rail-mounted gantry crane
ro-ro	roll-on-roll-off – ferry-type vessel onto which goods and containers can be driven, usually via a ramp
RTG	rubber-tyred gantry crane
SATL	semi-automatic twistlock
SIGTTO	Society of International Gas Tankers and Terminal Operators
SOLAS	International Convention for the Safety of Life at Sea, 1974
sto-ro	a vessel with a capacity for break-bulk cargo, as well as vehicles or trailer-borne cargo, as in the forest products trade
SWL	safe working load
UN ECE	United Nations Economic Commission for Europe
UNEP	United Nations Environment Programme

1. Introduction, scope, implementation and definitions

1.1. Introduction

1.1.1. General overview of the port industry

1. The international port industry dates from the earliest days of civilization. Since that time it has developed steadily over the years. However, cargo-handling methods that were both arduous and dangerous remained largely unchanged until the introduction of containers and roll-on-roll-off ("ro-ro") systems in the 1960s. Technical developments have continued since then, including the introduction of increasingly sophisticated cargo-handling equipment with greatly increased capacity and reach. While many of these changes in cargo-handling methods have resulted in significant improvements for the safety of portworkers, some changes have introduced new hazards and port work is still regarded as an occupation with very high accident rates. Moreover, privatization in the industry has led to considerable changes in the organization of ports and the employment of people in them, including increased use of non-permanent workers. Fortunately, systems for identifying and managing risks have also been developed and the need for investment in the training and skills of portworkers has been increasingly recognized.

2. Each port needs to develop working practices that will safeguard the safety and health of portworkers in the light of its own specific circumstances; these can be based on guidelines, such as those included in this code of practice, and on the well-established general principles set out in the relevant

international labour (ILO) Conventions and Recommendations, other codes of practice and guidelines. A full list is given in the references at the end of this code.

1.1.2. Reasons for the publication of this code

1. This ILO code of practice, which supersedes *Safety and health in dock work*, is in line with the general spirit of the "Conclusions concerning ILO standards-related activities in the area of occupational safety and health – A global strategy", adopted by the 91st Session of the International Labour Conference in June 2003.[1] Paragraph 9 of the ILO action plan for the promotion of safety and health at work, which forms a part of the Conclusions, states: "Occupational safety and health is an area which is in constant technical evolution. High-level instruments to be developed should therefore focus on key principles. Requirements that are more subject to obsolescence should be addressed through detailed guidance in the form of codes of practice and technical guidelines. The ILO should develop a methodology for a systematic updating of such codes and guidelines." This code seeks to give effect to these Conclusions.

2. The first edition of *Safety and health in dock work* was published in 1958 and complemented the Protection against Accidents (Dockers) Convention (Revised), 1932 (No. 32), which had replaced the earlier 1929 Convention. In 1976, a separate volume, *Guide to safety and health in dock work*, was published as a complement to the code of practice. A second, updated edition of the code was pub-

---

[1] ILO: "Report of the Committee on Occupational Safety and Health", *Provisional Record* No. 22, pp. 40-44, *Record of Proceedings*, Vol. II, International Labour Conference, 91st Session, Geneva, 2003.

lished in 1977 to take into account developments in the industry during the preceding 20 years. In 1979, Convention No. 32 was revised by the adoption of the Occupational Safety and Health (Dock Work) Convention (No. 152), and Recommendation (No. 160). Since 1979, Convention No. 152 has been ratified by a number of countries and used as the basis for legislation relating to port work in many others.

3. The second edition of the code, and the guide, did not reflect the requirements of Convention No. 152 and Recommendation No. 160, as they had been written earlier. Furthermore, technical developments had continued to take place in the port industry and some of the advice contained in these two documents had become obsolete. In addition, there was little advice on health matters, despite the considerably increased attention that has rightly been given to such matters in recent years. Accordingly, it was considered that the time had come to revise the code and the guide and combine them in one publication, to make it easier to use and to help in the implementation of the Convention and its complementary Recommendation. It is hoped that the availability of *Safety and health in ports* will help to raise the profile of safety and health in ports in all parts of the world and encourage more countries to ratify Convention No. 152, or otherwise implement its provisions. The Governing Body decided that Convention No. 152 is up to date and should be promoted. [2]

---

[2] See GB.270/LILS/WP/PRS/1/2, paras. 128-134.

1.2. Scope

1. The scope of this code reflects that of Convention No. 152 and Recommendation No. 160. It covers all aspects of work in ports where goods or passengers are loaded onto or unloaded from ships, including work incidental to such loading or unloading activities in the port area. It is not limited to international trade and is equally applicable to domestic operations, including those on inland waterways.

2. The final part of the code gives some brief guidance on matters that are not directly covered by Convention No. 152, but are nevertheless essential to the safe and proper operation of a port.

3. This code generally includes the material that was included in the previous code and the guide, with a few exceptions. General guidance on offices and workshops has been omitted, while guidelines on warehouses and fire precautions in ports have been shortened, since the precautions to be taken and the standards to be achieved are no different from those in any other industry, in accordance with national provisions. The chapter on nuclear-powered merchant vessels included in the guide has also been omitted. When that chapter was written, it was expected that the number of such ships would increase considerably. This has not happened.

4. Although in many ports certain working practices have been replaced by newer methods, older conventional methods continue to be used in other ports and limited advice on such methods has been retained in this code.

5. A very wide range of different cargo-handling activities is carried out in ports. It is not practical to cover all of

them in detail in one volume. However, this code is intended to cover the most common activities. Where appropriate, reference is made to other international publications.

6. It is necessary to take special additional precautions in connection with the loading and unloading of goods onto or from ships at offshore oil installations. These are beyond the direct scope of this code, but much of the guidance contained in it will be relevant to such operations.

1.3. Implementation

1. This code is intended to be a concise set of recommendations based on good practice in the industry. The advice should be useful to all bodies and persons concerned with safety and health in port work. These include government authorities, employers, workers and their representatives, manufacturers and suppliers of equipment, and professional bodies dealing with occupational safety and health.

2. It is appreciated that not all the provisions of the code will necessarily be applicable as they stand to all countries and all regions. In some cases, provisions may need to be adapted to local conditions.

3. Although much of the guidance in the code reflects long-established standards in the industry, advice on various matters has been updated to reflect modern standards in ports and other industries. The most significant of these relate to lighting and the height of fencing.

4. It is recognized that, although there should be no difficulty in applying the revised recommendations in the code to new equipment, there may be difficulties in applying

some of them to existing equipment. Wherever it is practical to do so, such equipment should be brought into compliance with the code as soon as practicable, for example during a major refit or replacement.

5. It is recognized that employers, workers or their representatives should cooperate and consult each other in respect to safety and health matters. Subject to this, the code should encourage employers and workers by means of cooperation and consultation for continuous improvement in safety and health levels.

1.4. Innovations in ports

Before technological or other innovations, and/or new work practices involving such innovations, are introduced in ports that may impact on safety and health of portworkers, the following should apply:

- It has been well established on the basis of evidence and data that the new operations can be done in a safe and proper manner and that safe working conditions are maintained.
- Consultations on safety and health aspects have taken place between employers and workers and their representatives, and agreement on these matters has been reached between them on the introduction of the innovations in question, with the involvement, as appropriate, of the competent authority of the State.
- Mechanisms have been established for monitoring the safe use of any technology; such monitoring should involve employers and workers, and their representatives.
- Relevant national laws and regulations, as well as all safety and health standards, have been complied with

and the guidance in this ILO code of practice should be taken into account.

1.5. Definitions

1. The definitions of the terms in this paragraph are those in Convention No. 152. As such, they apply throughout this code:

— *Port work* covers all and any of the part of the work of loading or unloading of any ship, as well as any work incidental thereto.

— *Access* includes egress.

— *Authorized person* – A person authorized by the employer, the master of the ship or a responsible person to undertake a specific task or tasks, and possessing the necessary technical knowledge and experience.

— *Competent person* – A person possessing the knowledge and experience required for the performance of a specific duty or duties and acceptable as such to the competent authority.

— *Lifting appliance* – Covers all stationary or mobile cargo-handling appliances, including shore-based power-operated ramps, used on shore or on board ship for suspending, raising or lowering loads or moving them from one position to another while suspended or supported.

— *Loose gear* – Covers any gear by means of which a load can be attached to a lifting appliance but which does not form an integral part of the appliance or load.

— *Responsible person* – A person appointed by the employer, the master of the ship or the owner of the gear, as the case may be, to be responsible for the performance of a specific duty or duties, and who has sufficient knowledge and experience and the requisite authority for the proper performance of the duty or duties.

— *Ship* – Covers any kind of ship, vessel, barge, lighter or hovercraft, excluding ships of war.

— *Portworker* – Any person engaged in port work.

2. The following definitions also apply for the purposes of this code unless otherwise stated:

— *Competent authority* means any minister, national regulatory body or other authority empowered to issue regulations, orders or other instructions having the force of law. Such authorities may include enforcing authorities such as government departments, local authorities or institutions, and port authorities.

— *Container* means a container as defined by the International Maritime Organization (IMO) in the International Convention for Safe Containers (CSC), 1972. Containers are rigid, rectangular, reusable cargo units intended for the intermodal road, rail or sea transport of packaged or bulk cargo by one or more means of transport without intermediate reloading. Containers may be general cargo containers, such as general purpose containers, open top, platform or platform-based containers, specific purpose containers such as tank containers, thermal containers or dry bulk containers, or named cargo-type containers. Most containers now in use are ISO series 1 freight containers. Requirements for their specification and testing are contained in the ISO 1496 *Series 1 freight containers – Specification and testing* family of standards. The term does not include a swap body, containers specifically designed for transport by air, any vehicle, cargo in a container, or the packaging of cargo; however, it does include containers when carried on a trailer or chassis.

— *Port area* means any port and surrounding area that is used for purposes incidental to the loading and unloading of passengers or cargo onto or from ships. In many cases, port areas may be defined by public or private legislation. Such areas may include factories or other enterprises unrelated to cargo-handling operations. This code is not intended to apply to the operation of such enterprises.

— *Explosion-protected* refers to electrical equipment that is constructed and installed in such a way that it is not liable to ignite a flammable or explosive atmosphere should it occur. Such equipment should be certified as complying with an appropriate standard acceptable to the competent authority.

— *Factor of safety* is the numerical value obtained by dividing the minimum breaking load or tension of an item of equipment by its certificated safe working load.

— *Heavy lift derrick* is a ship's derrick that is specially rigged for use from time to time in order to lift loads greater than those that may be lifted by the ship's light or general purpose lifting gear.

— *Legal requirements* are the requirements of any relevant international, national, local or port instruments, laws, by-laws, regulations or rules.

— *Ship's derrick crane* refers to a ship's derrick having a boom which may be raised, lowered and slewed transversely while supporting a load, by means of winches which either form an integral part of the arrangement or are used primarily with it.

— *Skeletal trailer* is a chassis used for moving containers, the longitudinal members of the chassis consisting typically of one or two longitudinal beams that are fitted at or near their end with transverse members to which the wheels and corner fittings are attached.

— *Transporter* is a rail-mounted or rubber-tyred gantry crane equipped with a horizontal bridge between its legs from which is operated a trolley or trolleys used with such items of equipment as grabs, magnets and container spreaders. Transporters are capable of straddling several rows of containers.

3. The following definitions apply for the purposes of Chapters 4 to 7:

— *In-service* describes a lifting appliance when handling loads up to its safe working loads in permissible wind speeds and other conditions specified by the manufacturer.

— *Inspection* refers to a visual inspection by a responsible person carried out in order to determine whether, in so far as can be ascertained in such manner, the equipment is safe for continued use.

— *Limiting device* is a device that automatically stops a lifting appliance motion or function when it reaches a prescribed limit (including limit or microswitches).

— *Out of service* means that the lifting appliance is without load on the load-lifting attachment and is either not required for use or is out of use under conditions specified by the manufacturer.

— *Personnel carrier* means a device that is attached to a lifting appliance for the purpose of lifting people.

— *Safe working load limiter* is a device that automatically prevents a lifting appliance from handling loads that exceed its safe working load by more than a specified amount.

— *Safe working load indicator* is a device that automatically provides acoustic and/or visual warnings when the load on a lifting appliance approaches or exceeds the safe working load by a specified amount.

— *Radius indicator* is a device that automatically shows the current operating radius of a lifting appliance and indicates the safe working load corresponding to that radius.

— *Safe working load* is the maximum gross load that may be safely lifted by a lifting appliance or item of loose gear in a given condition (sometimes referred to as "rated load" or "working load limit").

— *Thorough examination* means a detailed visual examination by a competent person, supplemented if necessary by other suitable means or measures, in order to arrive at a reliable conclusion as to the safety of the item of equipment examined.

4. Other terms used in this code are defined in the particular section to which they relate.

2. General provisions

2.1. Responsibilities

2.1.1. General requirements

Safety in ports is the responsibility of *everyone* who is directly or indirectly concerned with work in ports and needs to cooperate to develop safe systems of work and ensure that they are put into practice. The introduction of new ideas and concepts in cargo handling demands that special attention be paid to safety requirements. The guidance given in this code of practice relates to both new and existing working practices.

2.1.2. Competent authorities

1. When more than one authority is responsible for drawing up relevant legal requirements that apply to ports, it is essential that they liaise in order to ensure that their requirements are consistent with the relevant international instruments.

2. Competent authorities should ensure that legal requirements on safety and health in ports are put into practice. These should clearly define the bodies responsible for enforcing them and clearly identify duty holders. Enforcement bodies should also carry out accident and injury prevention activities, including the provision of appropriate information.

3. It is highly desirable that occupational safety and health regulations in each country be based on relevant international texts, including instruments adopted by the International Labour Organization (ILO), the International

Maritime Organization (IMO) and the International Organization for Standardization (ISO). The legal requirements relating to port work should implement the provisions of Convention No. 152 and should apply to ships of all flags when in a port.

4. Legal requirements should be framed in goal-setting terms, specifying the objectives to be achieved, rather than being prescriptive, thus allowing flexibility in the methods of achieving the objectives. This code will help competent authorities to publish guidance on how the objectives of their legal requirements based on Convention No. 152 can be achieved.

2.1.3. Port employers

1. Port employers, port authorities who carry out the function of a port employer and any other person who employs workers, on a permanent or temporary basis, should provide and maintain the workplace, plant and equipment they own, control or operate in a safe condition. They should also provide up-to-date written information on their safe use and operation.

2. Bodies employing or managing portworkers should:
— ensure that all portworkers (especially newly engaged workers) are properly instructed in the hazards of their respective occupations and the precautions that are necessary to avoid accidents and injuries;
— ensure that portworkers are appropriately informed of national or local legal requirements relating to their protection;
— provide appropriate supervision to ensure that the conditions of work of portworkers are as safe and healthy as

possible, and that the relevant safe systems of work are followed;

— if they are not themselves carrying out the port work, co-operate with those bodies and persons that are carrying it out in order to keep workplaces and plant and equipment safe.

2.1.4. Contractors and labour or service providers

Contractors and labour or service providers should co-operate with port authorities and other bodies working in port areas to protect the safety and health of all persons who may be affected by their activities. In particular, they should ensure that:

— all workers they employ or supply are appropriately trained and competent to perform the work they are required to do in port areas;

— all such persons are aware of the particular hazards of the port areas in which they are to work, the hazards and precautions to be taken in connection with port work in general, and any local rules;

— all such persons are appropriately supervised;

— all plant and equipment that they supply or use is of sound construction and properly maintained in a safe condition;

— they supply such information as is necessary to others who may be affected by their activities;

— they cooperate with the port authority, other employers and any other relevant bodies.

2.1.5. Ships' officers

Ships' officers should cooperate with shore personnel as necessary. This should include:

— providing safe means of access to the ship and to any place on the ship where portworkers need to work;

— ensuring that any ship's equipment that will be used by portworkers is of sound construction and properly maintained;

— providing such information as is necessary to portworkers on the ship;

— ensuring that the activities of the ship's crew do not give rise to hazards to safety or health on the ship;

— ensuring that if the crew work together with portworkers, joint safe systems of work are followed to protect the safety and health of all involved.

2.1.6. Management

1. Management bears the prime responsibility for safety and health in port work. Managers should be given the necessary authority, resources, training and support to put the policy of senior management into practice.

2. The management of port authorities or shipping companies that provide cargo-handling equipment for use in port work should strictly observe the relevant legal requirements and should be responsible for the safety of the equipment which they provide.

3. Management should provide appropriate information about occupational safety and health (OSH) and offer relevant vocational training to workers.

4. Management should draw up, organize, implement and monitor appropriate systems of work (this should include the choice of equipment) such that the safety and health of portworkers is ensured. Management should participate in the vocational training of all personnel.

5. Management should ensure that appropriate corrective action is taken as soon as any faults in plant or equipment or any hazards are reported to them, such action including the stoppage of workplaces or equipment if necessary.

6. Management should ensure that the work environment is safe at all times. Managers should continually check that portworkers work in accordance with the agreed safe systems of work and that any hazards identified are controlled.

7. It is essential that management comply with its own rules at all times, in order to be in a stronger position to demand such compliance from the workers.

2.1.7. Supervisors

1. Supervisors are an important level of management and the guidance relating to management also applies to them. However, supervisors occupy a particularly important position in the management chain as the representatives of management with whom portworkers are most likely to come into contact and by whom they are most likely to be influenced.

2. It is imperative that supervisors are fully aware of the need for port work to be carried out safely as well as efficiently, and they should know how work is carried out in practice.

3. In particular, supervisors should:

— bring to the attention of the portworkers the hazards and risks of the work they carry out and the need to follow safe systems of work;

— ensure that portworkers are provided with and use when necessary the correct protective clothing and equipment, which should be issued and maintained in good and efficient working condition;

— ensure that any defects in plant or procedures or other hazards, which they see or are reported to them, are dealt with promptly.

2.1.8. Portworkers

1. Safety is also a matter for all portworkers, who should:

— inform themselves of the risks inherent in their work and take full advantage of any vocational training courses available;

— cooperate with ships' officers to ensure that their activities do not give rise to hazards to the safety or health of crew members;

— acquaint themselves with and carry out all safety and health instructions relating to their work;

— comply strictly with all safety rules and instructions at all times;

— make proper use of all safeguards, safety devices and other appliances furnished for their protection or the protection of others;

— refrain from careless or reckless practices or actions that are likely to result in accidents or injuries to health;

— as soon as practicable, notify their supervisor (and, where appropriate, their trade union or a competent authority inspector) of any operation or equipment which they consider to be defective or otherwise dangerous. Such operations or equipment should not be further used until it has been checked and approved for further use;

— cooperate in the training of new workers, giving them the benefit of their experience;

— not interfere with, remove, alter or displace any safety devices and other appliances provided for their personal protection or that of others, or interfere with any procedure or safe system of work, except in an emergency or with proper authorization;

— be aware that other persons might be affected by their actions when carrying out port work. In some countries, portworkers have a legal responsibility in connection with the safety and health of others, as well as themselves.

2. Workers' organizations should be considered active partners in the development and administration of OSH material and courses for portworkers. To this end, they should also be provided with adequate means and assistance, as agreed by employers. In some cases, information given to portworkers by their own unions can have a much greater effect than information from other sources.

2.1.9. Self-employed persons

1. Self-employed persons are responsible for the safety and health of themselves and others who may be affected by their actions.

2. Self-employed persons should ensure that they are appropriately trained and competent to carry out their work, and should do that work in accordance with local legal requirements. It is essential that they liaise and cooperate with port authorities, employers and other relevant bodies, as appropriate.

2.1.10. Safety and health advisers

1. Safety and health advisers should assist management in implementing its policy to provide a safe and healthy workplace, and should give advice on safety and health matters to managers, supervisors and workers and their representatives, as appropriate.

2. The work of a safety and health adviser may include involvement in:
— updating or replacement of existing equipment;
— analysis of accident rates and trends;
— presenting the results of the analyses to management and workers and their representatives, as appropriate;
— development and revision of safe systems of work;
— investigation of accidents;
— proposals for new projects;
— safety audits;
— safety committees;
— training.

2.1.11. Other persons at work

Any other persons who may be present at work in port areas in addition to persons who carry out cargo-handling operations (e.g. hauliers, ships' crew members, pilots, ships' agents, immigration and customs officers, inspectors, members of the emergency services) should cooperate with the management of the port authority and other organizations with which they are working, and should comply with all relevant legal requirements.

2.1.12. Passengers and other non-workers

Passengers and other members of the general public who may be present in port areas but do not carry out port work should be separated from hazardous operations and instructed on the actions they should take by means of notices, verbal instructions or otherwise, and should comply with such instructions.

2.2. Management of safety and health

2.2.1. General requirements

1. The resources necessary to safeguard the safety and health of all persons affected by port operations should be managed so that a balance is achieved between the risks of operations and the cost of eliminating or reducing accidents. The real costs of injuries and ill health and the risks from the hazards of operations should therefore be assessed.

2. The true financial costs of accidents and illness should include the cost of direct damage, lost time and personal injury claims, as well as consequential costs such as time spent in administration, defending any claims that might be made, and replacing workers. The costs of accidents that do not result in injury should not be overlooked; they can provide an effective warning of potentially more serious incidents in future, thus saving considerable sums.

3. The outcome of an event may range from no injury to fatal injury and major damage, with only the smallest change to one factor. A "total loss" approach to accident prevention recognizes this fact and includes investigation of non-injury incidents. Organizations need to learn from all such incidents in order to achieve effective control.

2.2.2. Risk assessment systems

1. The difference between "hazard" and "risk" should be clearly understood:

— A *hazard* is a source of potential harm or damage, and may be a physical item or situation.

— A *risk* is the combination of the likelihood and the consequence of a specific hazard.

2. Risk assessment is an essential part of safety management. It provides a sound basis for the improvement of safety. It should cover all work tasks and hazards in the workplace and allows hazards to be assessed to see how harmful they are.

3. A risk-based safety management system requires management personnel to identify which activities need to be controlled within their organization and to interlink those activities for effective management. A risk-based approach allows for a continual improvement of standards, whereas a quality-based system merely requires adherence to a fixed standard.

4. Risk-assessment systems may be qualitative or quantitative. In qualitative risk assessment, risk is estimated by methods such as task analysis, identification of human factors and performance modelling. In quantitative risk assessment, risk is estimated by taking into account the probability

and severity of the outcome of a hazard. This is the method most commonly used to assess the risk of hazards in ports.

5. In its simplest form, the quantitative risk rating is the product of the probability of a hazard occurring and the potential consequences, including their severity (see table below).

6. These two factors should be determined independently. Although a potential consequence may be extremely serious, the probability of it occurring may be very low.

7. Multiplying these two factors gives a range of risk ratings between 1 and 25. These can indicate high-risk situations (20-25) that require rapid action, medium-risk situations (10-16) that require action or further evaluation within an appropriate period, and low-risk situations (1-9) that may require relatively little or no action.

8. More detailed risk-assessment systems also consider the frequency of the presence of the hazard.

9. Quantitative risk assessment is not a precise science but a tool to assist decision-making. It should not be used as a substitute for common sense when a hazard is patently obvious.

Hazard probability		Hazard severity	
Very likely	5	Very high	5
Likely	4	High	4
Quite possible	3	Moderate	3
Possible	2	Slight	2
Not likely	1	Nil	1

10. Risk assessment is best undertaken by a team including:

— a responsible manager;

— a supervisor;

— a worker representative;

— a safety adviser;

— a health adviser, where appropriate.

11. Any action found to be necessary should be planned and implemented within an agreed time scale. It should be checked that the action has been taken.

2.2.3. Safety and health management systems

1. National and local safety and health management systems for ports should be based on risk assessment, in accordance with the main elements of the ILO's *Guidelines on occupational safety and health management systems, ILO-OSH 2001.* These are:

— *Policy*. A clear statement of the organization's policy for safety and health involving workers at all levels.

— *Organization*. Specification of responsibilities and accountability, and necessary competencies and training requirements. These should be fully documented and effectively communicated to all concerned.

— *Planning*. Planning of development and implementation of the management system based on the latest review. This should identify measures necessary to eliminate or control hazards and set realistic objectives for the current period.

— *Evaluation*. Monitoring and measurement of current performance, investigation of accidents, periodic audits and review of the management system.
— *Action*. The implementation of necessary action to achieve continuous improvement of occupational safety and health.

2.3. Safe systems of work

1. Accidents are unplanned events. Working in a structured manner that recognizes and controls potential hazards can minimize such events. This is the basis of a safe system of work. Such systems result in safer and more efficient operations. Although they may not have been developed with safety in mind, quality control systems similarly result in safer operations by ensuring that operations follow specified patterns, thereby minimizing unplanned events.

2. Development of safe systems of work should include consideration of the:

— operations to be performed;
— workers who will carry them out;
— location of the work;
— working environment;
— nature of the cargo to be handled;
— plant, equipment and materials to be used;
— precautions to be taken, including any necessary emergency arrangements.

3. A safe system of work should specify:

— the task;
— necessary competencies of workers;

— equipment to be used, including protective equipment where necessary;

— potential hazards;

— control of the relevant hazards;

— procedures to be followed;

— control and supervision.

4. To be effective, a safe system of work should be developed in consultation with all parties involved with putting it into practice. Once finalized, it should be promulgated by appropriate means and any necessary training carried out before it is put into effect. Supervisory staff should monitor the implementation and effectiveness of the system in practice and be alert for any problems that may occur.

5. Safe systems of work should be reviewed periodically in the light of changes and operational experience, and revised as necessary.

2.4. Organization

2.4.1. Organization for safety and health in ports

1. The close interdependence of productivity and safety and health at work should be recognized by all who work in ports.

2. A permanent service responsible for OSH should be established in each port. The service should have the following tasks:

— promotion of OSH throughout the port and prevention of occupational accidents and diseases;

— inspection of workplaces;

— investigation into the causes of accidents that lead to death, serious injury or serious material damage;

— informing management and portworkers of such accidents and the lessons to be learned from them;

— informing management of incidents involving non-compliance with safety regulations;

— making formal reports on breaches of legal requirements;

— where necessary, drawing the attention of the relevant competent authority to urgent cases in which its immediate action or advice may be required;

— at regular intervals, drawing up reports of relevant activities, including accident statistics and practical advice on safety and health.

3. The port authority, even if not directly involved in port operations, should have the overall control of the operation of safe systems of work, the promotion of a safety culture and the development of safety and health in the port. The port authority should set up a central port safety and health committee with the help of employers and portworkers for fostering the necessary cooperation between all bodies involved in port work.

2.4.2. Safety and health committees

1. The composition and functions of a port safety and health committee depend on the number of employers and workers in the port and the organization of work in it. Particular attention should be paid to the problems associated with the increasing mobility of labour and the use of contract or other non-permanent workers in ports. Normally, the committee is composed of representatives of manage-

ment and workers in equal numbers. Committees can be set up to cover an entire port, as well as individual facilities or enterprises.

2. The functions of safety and health committees include:

— drawing up rules setting out safe systems of work for operations, and revising them when necessary;
— consideration of all proposals submitted for the improvement of working methods in the interests of safety;
— consideration of reports of inquiries into accidents and the drawing of lessons from them with a view to preventing recurrence;
— dissemination of information to portworkers and employers about hazards inherent in the work and ways of eliminating them or protecting themselves against them. This may include the preparation of safety leaflets, posters, and so on.

3. To perform these functions, the committee should be kept regularly informed of all dangerous incidents, accidents and occupational diseases that occur. It should also be informed of dangerous and unhealthy working conditions that are found, before they result in an accident or an incident of ill health.

4. In addition to a port safety and health committee, separate committees, affiliated to the port committee, should be set up by individual employers. A representative of the port authority should attend some meetings of these committees.

5. In some countries there is also a national port health and safety committee consisting of representatives of the

relevant competent authorities, port employers and port-workers. This permits the systematic supply of information to all interested parties and enables them to learn from relevant incidents and experiences in many ports without identifying them. These committees have been found to be of considerable mutual benefit by all who have taken part in them.

6. The safety and health committee should cooperate with the port security committee as appropriate.

7. In the event of a conflict, safety and health should be paramount.

2.4.3. Safety representatives

1. The function of safety representatives is to represent other workers. In some countries, they may be appointed in accordance with national legal requirements and have a number of additional functions.

2. Safety representatives may be appointed by recognized trade unions or, where they do not exist, by groups of workers.

3. Safety representatives should take an active part in the work of safety and health committees by providing a channel for the flow of information both from workers to management and from management to workers. Such information should not be restricted to a particular group of workers but disseminated to all relevant workers.

4. Safety representatives should be encouraged to raise safety and health concerns with the committee. However, they should not "save up" reports of defective equipment or other obvious hazards that need attention for the committee meeting. Such reports should be brought to the attention of

the appropriate person and dealt with accordingly. They should only be raised with the committee if appropriate action has not been taken or when other members of the committee can benefit from lessons that can be learned from the report.

2.5. Reporting and investigation of accidents

2.5.1. In-house reporting of accidents

1. All cases of injury or occupational illness to portworkers should be reported to the appropriate person in accordance with in-house instructions and procedures. Every effort should be made to encourage individuals to report incidents without repercussions because the report has been made.

2. Management should ensure that a written record is kept of all accidents, incidents of occupational illness and other occurrences, in accordance with national legal requirements.

3. Management should also identify other types of accidents that should be reported to them. These may be incidents resulting in certain damage to plant or property, or that have the potential to cause significant injury or damage (often termed "near misses").

2.5.2. Statutory reporting of accidents

1. All occupational accidents to portworkers causing loss of life, serious personal injury or incapacity for work, and incidents of specified occupational diseases, should be reported promptly to the relevant competent authority, in accordance with national legal requirements.

2. Certain other accidents, often termed "dangerous occurrences", may also be required to be reported to the competent authority, whether or not they have resulted in injury. These, and relevant occupational diseases, are specified in national legal requirements or by the competent authority. Examples may include the collapse of cranes or derricks, explosions and serious fires.

3. The competent authority should undertake an investigation into the causes and circumstances of any fatality and serious accident in accordance with national policy.

4. The scene of a fatal accident should, as far as practicable, be left undisturbed until it has been visited by a representative of the competent authority.

5. After a dangerous failure of plant or gear, the plant or gear should, as far as practicable, be kept available for inspection by the competent authority.

2.5.3. Investigation of accidents

1. All accidents should be investigated with a view to determining their cause and to determining the action that should be taken to prevent any similar accident in the future.

2. The formality and depth of the investigation should be proportional to the severity or potential severity of the accident. Often, only a minor change in circumstances can make the difference between an accident resulting in no injury, very minor injury, or a fatality. It should not be necessary to wait for a serious injury to occur before appropriate steps are taken to control a hazard. Action taken after a "near miss" can prevent future injuries and losses resulting from damage.

3. The names of witnesses should be recorded and any relevant photographs taken should be identified, captioned and dated.

4. The investigation should consider all the relevant evidence. This may include the site where the incident occurs, plant, the type of cargo being handled or substances being used, systems of work, responsibilities and people involved, including their physical or mental condition, training and competencies.

5. It is important to investigate not only the *direct* cause of an accident, but also to determine the *underlying* cause or causes, which are often the real cause of an accident. Human factors have been found to be relevant to a high proportion of accidents.

2.6. Selection and training

2.6.1. Selection of portworkers

1. The provisions of the ILO Discrimination (Employment and Occupation) Convention, 1958 (No. 111), should be fully applied in the selection process in order to eliminate any possible discrimination.

2. Portworkers should only be engaged following an appropriate selection process.

3. A portworker needs to have a good physical constitution. Normal reflexes and good eyesight are essential for safe port work, especially for operators of mechanical equipment.

4. Further training is necessary after selection if the worker is not already appropriately trained and fully competent for the type of work to which he or she is to be assigned.

5. Portworkers should be able to work in teams. It should be recognized that:

— their acts and behaviour will have a direct influence on those of other portworkers;

— the safety and output of other workers will depend on their own vigilance, work and knowledge;

— the safety of others may be dependent on their own safety.

6. Modern selection methods enable selection to be based on objective tests of medical fitness and aptitude, as appropriate. It is desirable for port occupational health staff and personnel officers, as well as relevant operational management, to be involved in the selection of portworkers.

7. No person under 18 years of age should be employed in port work, except as permitted by the competent authority in accordance with national legal requirements. However, workers over 16 years of age may be employed under an apprenticeship or other training scheme, subject to conditions prescribed by the national authority.

8. The main resources necessary for training portworkers are suitable premises and equipment, training material, teaching staff and appropriate remuneration of trainees.

2.6.2. Training needs

1. All portworkers should be trained to develop the knowledge, psychomotor and attitude skills which they need to enable them to do their work safely and efficiently, as well as to develop general safety awareness. Portworkers should be aware of the potential effects of their actions on others, as

well as the specific hazards of their work and methods to control them. Training should include both general induction training and training relevant to their specific work.

2. Consideration should be given to the need for continuation or refresher training in addition to initial training. This may be necessary to deal with technological advances and the introduction of new plant or working practices. It may also be necessary to eradicate bad practices that have developed with time and to remind workers of basic principles.

3. Records should be maintained of the training that each portworker has received and the competencies that have been attained.

2.6.3. Induction training

1. General induction training should be given to all persons who are to work in ports. This training should cover the general hazards associated with ports, which are often quite different from those encountered in other industries.

2. The training should also include site-specific hazards and relevant local rules, emergency arrangements and the need to cooperate with other persons working in the port. It can be accompanied by a leaflet for all visitors to a port setting out basic information, including the action to be taken in an emergency.

3. Persons with previous relevant training may not need to be required to repeat the general part of the induction training; however, they should receive the relevant site-specific information in all cases.

2.6.4. Job-specific training

1. The need for structured training, rather than just working alongside a more experienced worker, has increased following the mechanization of port work.

2. Job-specific training, including knowledge of cargo-handling methods, should be provided for all portworkers working with cargo, not just those operating plant or other equipment.

3. National vocational qualifications in relevant port work competencies should be developed. The holding of a relevant certificate of competency can assist potential employers in the selection of portworkers. However, the holding of such a certificate should not relieve an employer from the duty to check that the necessary skills have been maintained and can be put into practice.

4. Port authorities and private companies should be aware of the ILO Portworker Development Programme (PDP). This has been developed to give international guidance on vocational training for portworkers.

2.6.5. Training methods

Training should generally include both theoretical training in a classroom and practical training. Training of portworkers should normally be carried out in ports to allow demonstrations of working practices by experienced personnel and to bring trainees into close contact with future workplaces.

2.6.6. Evaluation of training

Attendance at a training course does not guarantee that a trainee has gained the necessary skills. Where appropriate, training courses should conclude with a suitable test that demonstrates that the trainee has reached the necessary level of skills. Successful trainees should be given a certificate specifying the skills and, where appropriate, the level attained.

2.7. Information for portworkers

1. Relevant information on matters that are likely to affect their safety or health should be available to all portworkers. The information should be given to them in writing or made available to them by other means.

2. The information should include relevant safe systems of work, material safety data sheets relating to dangerous cargo that they handle or dangerous substances that they use during their work, and reference to relevant port safety and health legal requirements.

2.8. Special facilities for disabled persons

Special facilities for disabled persons relating to safety and health should be provided as appropriate according to national legal requirements.

3. Port infrastructure, plant and equipment

3.1. General provisions

3.1.1. Separation of people and vehicles

With the mechanization of cargo-handling operations, the design, layout and maintenance of port infrastructure and plant and equipment have become increasingly important. As vehicles and mobile plant are now one of the main elements in fatal and serious accidents in ports, people should be separated from vehicles whenever this is practicable.

3.1.2. Surfaces

1. The surface of port areas should be:
— of adequate strength to support the heaviest loads that will be imposed on them;
— level, or with only a slight slope;
— free from holes, cracks, depressions, unnecessary kerbs or other raised objects;
— continuous;
— skid resistant.

2. The possible need for future repair should be considered when selecting surface materials.

3. As asphalt can be damaged by oil, fuel and other solvents, spillages should be cleaned up immediately to prevent or minimize damage.

4. Plain metal surfaces, such as those on brows or ramps, can become slippery, particularly when wet. The use of chequer plate or other plates with raised patterns or non-slip coatings should be considered.

5. Wooden structures should be built of wood that is suitable for use at the location in question. Additional protection may be provided by the ███████████ preservatives. Wood should not be covered w██████████████ther materials that will hide its condition and █████████████celerated hidden rot or other deterioration. ████

6. Plastic surface coverings can include a variety of non-slip finishes.

7. All surfaces other than ramps, etc., should be as level as reasonably practicable while providing adequate drainage. Any slope on quays or other operational areas should not exceed 1 per cent and should not slope towards the edge of a quay. Drainage systems should include appropriate interceptors to prevent maritime pollution.

8. Ramps or slopes used by lift trucks or other cargo-handling vehicles should not have a gradient steeper than 1 in 10 unless the vehicles have been designed to operate safely on such a gradient.

3.1.3. Lighting

1. Adequate lighting of all working port areas should be provided during the hours of darkness and at times of reduced visibility.

2. Different levels of lighting may be appropriate in different areas.

3. On access routes for people, plant and vehicles, and in lorry parks and similar areas, the minimum level of illumination should not be less than 10 lux.

4. In operational areas where people and vehicles or plant work together, the minimum level of illumination should not be less than 50 lux.

5. Light meters should be able to read to an accuracy of 1 lux. Meters should have a wide angle of acceptance in order to minimize errors due to directionality or low sensitivity to differing types of light sources, or be provided with the relevant correction factors.

6. Light measurements should normally be taken in the horizontal plane 1 m above the ground or other working surface. Measurements at a lower level may be necessary where there are obstructions that might conceal a tripping hazard. The meter should not be oriented towards a light source.

7. Records should be kept of all lighting measurements. These should include the date, time, weather conditions, location and details of the lighting and light meter.

8. Higher levels of lighting may be required at particularly dangerous places, such as shore gangways, accommodation ladders, steps and other breaks in quays or where detailed work is necessary. Where a higher level of lighting is required only occasionally, it may be provided by mobile or portable equipment.

9. Lighting should be as uniform as practicable. Sharp differences in lighting levels should be avoided.

10. The choice and positioning of light sources and each installation should be planned individually.

11. Lamps emitting monochromatic light, such as sodium-vapour lamps, give a good light in foggy weather but distort colours and may lead to confusion. They should be confined to non-operational areas. In operational areas,

Figure 1.   Tall lamp standard for illuminating a large area

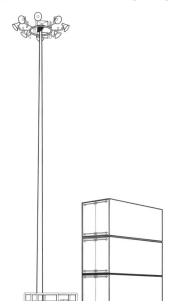

fluorescent or other lamps, which give a light more similar to daylight, should be used.

12. Tall lamp standards (over 12 m high) carrying several lamps can each illuminate a large area, cause less obstruction and reduce areas of shadow between containers (figure 1).

13. Lamp fittings should be provided with shades and diffusers to prevent light pollution and glare. Particular attention should be paid to preventing dazzle to small ships from reflection of light on water.

14. Lamp standards should be designed to allow the lamps to be cleaned and changed in safety.

15. At ports where operations do not take place for 24 hours per day, arrangements should be made to ensure that crews of ships berthed in the port have safe access to and from their ships. This may be ensured by the provision of sensors or switches on lamp standards on authorized walking routes that turn on lights for an appropriate period.

3.1.4. Fire precautions

3.1.4.1. General requirements

1. Fire precautions in ports should be provided in accordance with national legal requirements. These should consist of:

— fire protection;
— fire alarms;
— fire-fighting equipment;
— means of escape in case of fire.

2. Advice on fire-related matters should be obtained from fire authorities and insurance companies. Industry-specific advice may be available from appropriate industry organizations.

3.1.4.2. Fire protection

1. The principles of fire protection at ports are no different from those in other industries.

2. Whenever practicable, buildings and structures at ports should be constructed of non-combustible materials. Where this is not practicable, structures and construction materials that will reduce the probability of fire and limit the

consequences of any that do occur should be selected. National or local legal requirements generally set out standards for buildings or compartments (subdivisions of a building), particularly those where flammable or other dangerous substances are kept. Generally, fire separation walls should have a fire resistance of at least two hours.

3. Large buildings should be partitioned into fire-resistant compartments that do not exceed 9,000 m^2 in area.

4. Fire separation walls of a building or compartment should not have holes or gaps in them. Any doors that are necessary in such walls should be fire-resistant and self-closing. Spaces where services pass through fire separation walls should be fire protected.

5. Sources of ignition should be rigorously controlled, particularly in warehouses and other places where flammable materials are likely to be present.

3.1.4.3. Fire alarms

1. An effective fire alarm system should be provided throughout port areas. This may be by "break glass" fire alarm points or otherwise. If the system involves the use of a radio or telephone system, the system should operate at all times. Automatic systems can be arranged to sound alarms in relevant areas, alert the fire authority and operate automatic fire-extinguishing appliances, as appropriate.

2. In large premises, it may not be necessary to alert all persons in the port area immediately in the event of a fire, and a staged fire alarm system that allows different areas to be alerted may be appropriate. The fire alarm system in any building should be audible throughout the entire building.

3. The fire alarm system should be maintained in a fully operational condition at all times, particularly when maintenance work or alterations to premises are in progress.

3.1.4.4. Fire-fighting equipment

1. Appropriate means for fighting fire should be provided throughout port areas. These should include both portable first-aid fire extinguishers and fixed systems such as hoses and hydrants.

2. The location, type and number of fire-fighting equipment should be determined in accordance with national and local legal requirements.

3. Portable fire extinguishers should be grouped at clearly marked fire points. Fire points should be identified by clear and conspicuous signs or markings. These should be visible at all times and not obstructed by cargo or plant. If necessary, signs should be raised so that they can be seen above stored goods. Fire points should be located in such a way that the equipment can be brought into use quickly. Hydrants at warehouses should be close to doors.

4. All fire-fighting equipment and systems should be tested at regular intervals.

5. The choice of fire-fighting agent is determined by the type of fire that is likely to occur and the nature of materials that are likely to be involved. The use of an inappropriate fire-fighting agent can be extremely dangerous.

6. The most commonly used fire-fighting agents are:
— water;
— foam;

— carbon dioxide;

— powders.

7. Water is the most common fire-fighting agent and is suitable for use on most general fires. As well as extinguishing most fires, it also cools the surrounding area thus reducing the chance of the fire re-igniting or spreading.

8. The intake of a fixed fire main that takes water from a port should be below water at all states of the tide.

9. Hydrants should not be more than 80 m apart. International ship–shore connections to enable fire mains to be connected to those on ships (figure 2) should be available at all berths in accordance with IMO Resolution A.470(XII), and should conform to the dimensions in regulation II-2/19 of IMO International Convention for the Safety of Life at Sea (SOLAS), 1974.

10. Water pipes and hoses should be protected against collapse, impact by wheeled traffic or falling goods and frost.

11. Water and water-based foams should never be used to fight fires involving electrical equipment or chemicals that may react violently with it.

12. The shelf life of all chemicals used to make chemical foams should be determined and stocks renewed periodically.

13. Portable carbon dioxide extinguishers should not be used in confined or unventilated spaces. If a total flooding system is installed, it should provide an audible pre-warning of discharge in the protected space. This should be distinguishable from the fire alarm and give sufficient time for persons to escape before the discharge.

Figure 2.   Shore connection for international ship–shore fire
connection

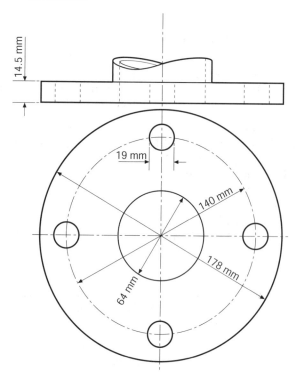

3.1.5. Means of escape in case of fire

1. Adequate means of escape in case of fire should be
provided from all places in ports and should lead to safe
places outside buildings.

2. Escape should normally be available by at least two different routes, except where very small travel distances are involved.

3. Fire assembly points to which persons can safely go in the event of a fire should be clearly identified.

4. Suitable access routes for emergency services in the event of fire should be provided throughout the premises. These should be clearly signed and kept clear of obstructions at all times.

3.2. Traffic routes

3.2.1. Roadways

1. Suitable roadways should provide safe access for vehicles to all parts of port areas.

2. Wherever practicable, vehicles and pedestrians should be separated.

3. The width of roadways should be suitable for the traffic that is likely to use them. This should take into account the width of vehicles and their loads, and their ability to manoeuvre. Traffic lanes should generally be at least 5 m wide. Under gantry quay cranes and in other restricted locations, a width of 7 m may be necessary to provide adequate clearances.

4. Lanes wide enough for road traffic should be provided between rail tracks and rubber-tyred gantries along quays.

5. A 2 m clearance should be left along quaysides to enable a 1 m unobstructed access for personnel to be available.

6. Roadways should be sited so as to allow the road ahead to be clearly visible for an appropriate distance. Sight lines should not be obstructed by corners of buildings, stacked goods or other obstructions.

7. Roadways should be separated from any fixed obstacles by a clearance of at least 900 mm.

8. Roadways should be unidirectional. Where this is not practicable, separation of the traffic streams by traffic cones, or otherwise, is desirable on all main traffic routes.

9. The edges of roadways should be clearly delineated by pavements or other clear markings. These should be clearly visible by both day and night. Yellow or white reflective road paint can be used for this purpose.

10. Particular attention should be paid to the design of any roundabouts (traffic circles) found to be necessary. Elongating the traffic island in the roundabout, rather than making it circular, can help to prevent overturning of vehicles.

11. Wherever practicable, the layout of roadways should be standardized throughout the port areas.

12. Provision for the safe parking of vehicles should be provided in appropriate places.

13. Appropriate warning signs should be provided on roadways in ports. These should conform to the national legal requirements for road signs. Wherever possible, standard international road signs should be used.

3.2.2. Walkways

1. Safe walkways should be provided to all parts of port areas to which persons with legitimate access have to walk. Such people include ships' crew members, pilots, passengers and contractors.

2. Wherever practicable, walkways should be separated from operational areas and vehicular traffic.

3. Walkways should be wide enough for the number of persons expected to use them at any one time.

4. The edges of walkways should be clearly delineated. The markings should be clearly visible by both day and night. Yellow or white road paint can be used for this purpose.

5. Markings to identify walkways should be clear and unambiguous (figure 3). There should be no doubt whether markings indicate a walkway, a plant crossing or other dangerous area, such as the track of plant.

6. Signs or markings should be provided at the ends of walkways and repeated at intervals along them as necessary.

7. Appropriate signs should be provided where walkways cross roadways.

8. International symbols and warning signs should be used whenever practicable. Pictorial symbols should be used on signs and on the surface of walkways to ensure that they are understood by users of all nationalities.

9. Obstructions in a walkway should be clearly marked or signed.

Figure 3.   Examples of markings and signs on walkways

3.2.3. Other matters

1. Cycle-ways to separate cyclists from vehicles and pedestrians should be provided in ports where there is significant cycle traffic.

2. All permanent or temporary obstructions should be clearly marked to be visible by day and night. Holes, lamp standards and other obstructions should be securely fenced as necessary.

3. Where a hole or trench has to be temporarily covered and is to be crossed by vehicles, the covering should:

— be of adequate strength;

— have a sufficient overlap at the edge of the hole or trench and be suitably anchored;

— be of sufficient width;

— be provided with a ramp where there is an abrupt difference of levels;

— have a notice at each end stating that vehicles should not attempt to pass other vehicles while on it.

4. If the covering is to be used by pedestrians, it should also:

— be fenced on either side; and

— have a notice at each end stating that pedestrians should give way to vehicles.

5. Overhead obstructions that may be struck by vehicles should be clearly marked. These may include overhead walkways and pipe bridges, and low doorways into buildings. Where necessary, notices specifying the maximum height of vehicles, including their loads, should be displayed (figure 4).

Figure 4.   Low doorway markings with maximum heights

3.3. Cargo-handling areas

3.3.1. Layout

1. Cargo-handling areas should be well surfaced and, where applicable, should comply with the provisions of sections 3.1.2 (Surfaces) and 3.1.3 (Lighting), above.

2. Lamp standards and similar structures that may necessarily be present should be protected with barriers against accidental damage by cargo-handling equipment and vehicles (figure 5).

3. The layout of cargo-handling areas should be such as to avoid the need for walkways to cross them, as far as is

Figure 5.   Barriers around lamp standard

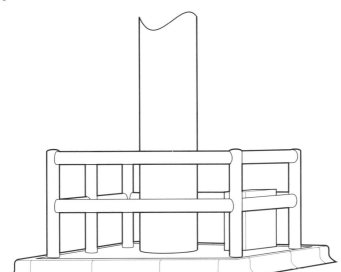

practicable. Any crossing points that are necessary should be marked and signed to warn both drivers and pedestrians of the potential presence of each other.

3.3.2. Edge protection

Secure fencing should generally be provided at all places from which a fall is likely to result in serious injury. This includes openings where there are sudden changes in level, such as the tops of steps, and open edges from which it is possible to fall more than 2 m or into water.

3.3.3. Quay edges

1. It is recognized that it is not practical to provide fencing along all the open edges of quays. Fencing should be provided at all dangerous corners and breaks in quay edges (figure 6), such as those at the sides of gangways, ramps or brows giving access to ships, pontoons or landing stages, walkways over lock gates or caissons, and the edges of quays overlooking open stretches of water.

2. Fencing should be provided at quay edges where a large number of passengers are likely to be present.

3. All edges of quays on which vehicles are used close to the edge should be protected by a continuous coping wall or robust rigid barrier of sufficient strength to prevent trucks and most other vehicles from accidentally falling into the water (figure 7). In general, the wall or barrier should be as high as practicable, but not less than 300 mm high. Highway-type barriers may be suitable for the purpose. On quays where only cars and other small vehicles are used, a lower

Figure 6.   Fencing of quay corners and breaks in quay edges

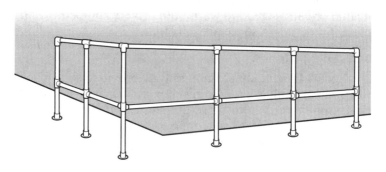

barrier may be sufficient but this should not be less than 200 mm high.

4. Gaps may be left in the wall or barrier where this is necessary to work capstans, use bollards or carry out other operations. The gaps should be no wider than is necessary and less than the width of a vehicle.

5. If vehicles are only very occasionally used near a quay edge, suitable temporary arrangements may be made. These may involve the provision of a temporary barrier or the positioning of a person to signal to the driver when a vehicle is operating close to the edge of the quay.

6. Where an existing rail-mounted crane passes close to the water's edge and it is not practical to provide fencing on the quay, it may be advisable to put a fixed handrail on the crane. This should not project in such a way as to dangerously reduce the clearance between the crane and the edge of the quay or any fencing that the crane may pass.

Figure 7.   Quay edge protection for vehicles

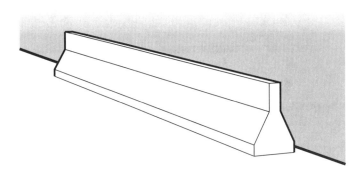

3.3.4. Fencing

1. All permanent fencing should be robustly constructed.

2. Fencing should generally consist of metal railings. Reinforced concrete barriers may be appropriate on structures alongside bodies of water and where there is heavy vehicular traffic for ro-ro or container operations.

3. Other fencing should depend on the nature of the hazard to be protected, the general layout of the immediate area and any nearby structures. Chains between stanchions provide only limited protection and should not be used as permanent fencing. Fencing should not stop immediately at the end of the danger zone but should continue a few metres beyond it.

4. The construction and location of fencing should allow for ships to be made fast and cast off easily.

5. Fencing should be at least 1 m high (figure 8). Metal railings should have a middle rail 500 mm above the quay

Figure 8.   Construction of fencing

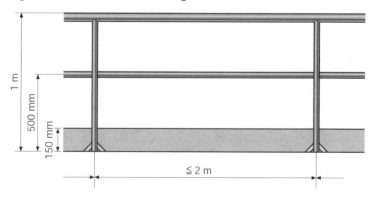

between stanchions that are placed not more than 2 m apart. Toe boards 150 mm high should be fitted where necessary.

6. Movable fencing may be used around temporary hazards and on the edges of berths. It can be removed while work is going on when necessary to avoid the fencing itself becoming a hazard.

7. Movable fencing should be used to protect stairs or steps at the water's edge or the edges of gangways where permanent fencing is not practicable.

3.3.5. Quayside ladders

1. Permanent ladders should be provided at the edge of any structure in a port from which persons may fall into deep water to enable them to climb out of the water (figure 9).

Figure 9.   Quayside ladder

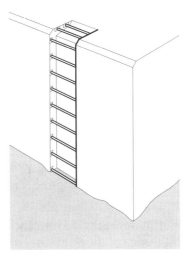

Such structures may include quays, jetties and dolphins and large mooring buoys.

2. Ladders should be spaced at intervals of not more than 50 m from each other or from steps.

3. Ladders should conform to section 3.5.3, where appropriate.

4. The bottom rung of ladders should be at least 1 m below the lowest level of the water at any time, or on the bed of the dock if there is less than 1 m of water at low tide.

5. Where the stringers of the ladder extend above the quayside, they should be opened out sufficiently to enable a person to pass through them, and should be sloped or curved in from the quay edge.

6. Where it is not practical for the ladder to extend at least 1 m above the top of the quayside, the stringers should extend as high as is practicable. Where no such extension is practicable, adequate handholds should be provided on the surface of the quayside in front of the ladder (figure 10). If these are recessed in the surface, the recess should be designed so that it allows drainage and does not fill with water or dirt. If handholds projecting above the surface of the quay and recesses are necessary, they should be clearly marked to draw attention to possible tripping hazards.

7. A permanent ladder should be protected on each side against damage from ships, unless it is recessed into the wall of the quay.

8. The top of a ladder recessed into a wall should never be obstructed by the edge of the quayside.

Figure 10. Handholds and drainage at top of quayside ladder

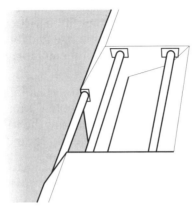

9. Ladders from the water should be conspicuous so as to be easily seen by anyone falling into the water. The tops of the ladder should be clearly visible to persons on the quayside.

10. At ports where it is not practicable to fit permanent ladders, or at quays that are used only occasionally and persons do not have to pass when no ship is berthed, temporary ladders should be provided and secured fore and aft of each ship loading or unloading.

3.3.6. Life-saving equipment

1. Adequate and suitable life-saving equipment should be provided and maintained for the rescue of anyone in danger of drowning.

2. Rescue equipment should consist of lifebuoys, throwing buoys or lines, grapnels, boathooks or other suitable equipment. Throwing-lines fitted to lifebuoys or similar equipment should be of suitable size and length, and should be made of polypropylene or other suitable material so that they will float.

3. Life-saving equipment should be located at suitable places at intervals of not more than 50 m. Such locations should be near the quay edge close to the tops of ladders or steps to the water, wherever practicable, and should include landing stages.

4. The equipment should be prominently mounted at a location painted in a conspicuous colour.

5. The locations should be kept free of obstructions so as to be easily visible at all times.

6. Lifebuoys and similar equipment should be hung up or contained in a case or cupboard of adequate size and conspicuous colour. Cases and cupboards can be arranged to alert a central point when opened or when the equipment is removed. This can aid calling the emergency services and reduce theft and vandalism.

7. Where theft and vandalism are serious problems, it may be satisfactory for life-saving equipment to be kept just inside nearby sheds or other buildings, provided that its location is clearly marked and it is immediately available at all times when work is in progress.

8. Suitable notices should be displayed with life-saving equipment giving clear instructions for raising the alarm in the event of an emergency and for the resuscitation of a person rescued from drowning.

9. Facilities to enable persons who have fallen into the water to support themselves while awaiting rescue should be provided between quayside ladders. The structure of the quay may be able to serve this purpose. More often chains are used. In enclosed quays, or where there is a very small tidal range, chains looped between fixed points may be provided. Where there is a large tidal range or in locks, vertical chains should be hung on the face of the quay. At least one such chain should be provided between adjacent quayside ladders.

3.4. Shore-side access to ships

3.4.1. General requirements

1. Port authorities and operators should provide safe access through port areas to and from ships. This should normally be by means of clearly signed, marked and lit walkways.

2. Port plans should be displayed at all entrances to port areas and elsewhere as necessary to enable crews and others to reach ships safely.

3. If ships berth at quays where pedestrians are prohibited, notices should be displayed at port entrances and alongside ships' gangways giving appropriate instructions. The notices should contain information about the necessary arrangements for obtaining safe transport through the port areas.

4. In some circumstances the shore side provides access to ships. Such means of access include passenger walkways and access towers to large ships. In all such cases the shore owners and operators should ensure that the equipment is of

good design and construction, properly installed and maintained in a safe condition.

3.4.2. Shore ramps and passenger walkways

1. Shore ramps may be necessary to provide access from shore to ships (figure 11), particularly ro-ro ships, or to floating landing stages and pontoons.

2. Shore ramps include all linkspans and brows (figure 12). Linkspans may include some form of lifting appliance that raises or lowers the roadway as necessary. Brows do not include any mechanically operated lifting appliance and are only used by pedestrians.

3. Passenger walkways should generally be separate from vehicle linkspans. If passenger walkways are combined with a vehicle linkspan, the walkways should be separated from the roadway, preferably by a robust fence.

4. All shore ramps and passenger walkways should be designed, constructed and installed in accordance with the relevant national legal requirements.

5. The design of shore ramps and passenger walkways should:

— cater for the maximum likely movement resulting from the tidal range, and ranging and drifting of the ship or pontoon at its moorings;

— cater for the maximum forces to which they may be subjected during mooring;

— recognize that they are likely to be subjected to forces in all three planes and torsional forces;

Figure 11.   Linkspans and walkways

— include safe access to all points to which access is necessary for routine maintenance and inspection;

— include transition flaps at their ends, as necessary.

    6.   Pontoons supporting shore ramps or passenger walkways should be designed to be as steady as practicable in all weather and tide conditions.

Figure 12.   Pontoon landing stage with linkspan or brow

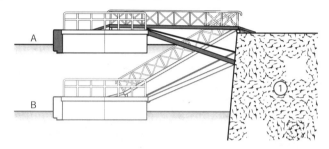

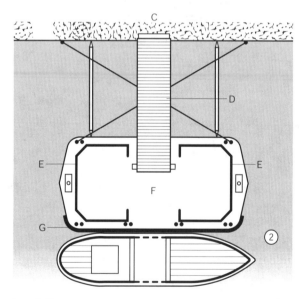

1. Elevation.   2. Plan.

A. High tide.   B. Low tide.   C. Embankment or quay.   D. Linkspan or brow.
E. Fencing.   F. Pontoon.   G. Fender.

3.4.3. Landing stages

1. Landing stages should be fenced by fixed railings (figure 13). A gap of not more than 2 m may be left in the railings to permit embarking and disembarking. Hinged or movable railings or chains should be provided to bridge this gap when it is not in use.

2. Landing stages should be provided with suitable bollards, cleats or other arrangements of adequate strength to which ships may be made fast.

3. Landing stages and other pontoons should be provided with hanging chains for the support of persons in the water.

Figure 13. Fencing of pontoon landing stage (for clarity, closure not shown)

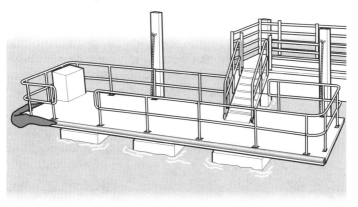

4. Appropriate lighting should be provided at all landing stages that are used during the hours of darkness (see section 3.1.3).

3.4.4. Steps and stairways

1. Steps and fixed stairways providing access from the water should extend at least 1 m below the lowest level of the water in the dock basin at any time or to the bed of the dock if there is less than 1 m of water at low tide.

2. Stairs and fixed stairways giving access to harbour craft should be equipped with wooden or rubber fenders. The gap between the side of the steps and the side of any craft should not exceed 300 mm unless a gangway is used for access.

3. Steps and fixed stairways should be constructed so as to minimize the accumulation of mud, dirt, marine growth or other matter that is likely to make them slippery. This can be achieved by the use of gratings or anti-slip finishes on treads.

4. Fixed handrails should be provided on the landward side of all steps and fixed stairways. Fencing (see section 3.3.4) should be provided where necessary on the water side of all fixed stairways (figure 14). Sections of such fencing may be removable if necessary.

5. Appropriate lighting should be provided at all steps and fixed stairways that are to be used during the hours of darkness. If the general port lighting is not adequate, additional lighting on the steps should be provided (see section 3.1.3).

Figure 14.   Moveable fencing on steps

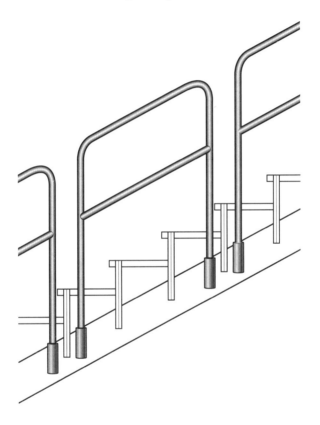

3.4.5. Quayside ladders

1. Quayside ladders (see section 3.3.5) may be used for access to small ships such as fishing vessels.

65

2. Where quayside ladders are regularly used for access to small ships, additional ladders may be necessary. Suitable bollards, mooring rings or other facilities should be provided to which ships may be made fast to prevent the ladders being used for mooring purposes.

3.5. Access to terminal buildings, structures and plant

3.5.1. General requirements

1. The requirement to provide safe means of access applies to all places to which persons in ports need to go in the course of their work. It includes permanent and temporary means of access to all parts of buildings, structures and mobile plant.

2. Permanent fixed access should be provided to all places to which persons have to go regularly, including the cabs of mobile plant.

3.5.2. Stairways and steps

1. All stairways and steps in buildings should comply with national legal requirements.

2. The open sides of stairways and steps should be protected by suitable fencing. This should at least have a top rail 1 m above the treads of the stairs and one intermediate rail.

3. A handrail should be provided on both sides of a stairway. This can be the top rail of the fencing. An additional intermediate handrail should be provided on stairways that are 2 m or more in width.

4. The treads of stairways should have a slip-resistant surface.

5. All stairways and steps should be maintained in a safe condition. Wear-resistant nosings that can easily be replaced when necessary should be used at the edges or noses of treads of stairways that are liable to considerable wear. Care should be taken to ensure that such nosings do not project significantly in front of steps or work loose to become a danger.

3.5.3. Fixed ladders and walkways

1. All fixed ladders should be of steel construction.

2. Rungs or treads of ladders should:

— be equally spaced at intervals of not less than 250 mm or more than 350 mm;

— provide a foothold not less than 150 mm deep and 350 mm wide;

— be horizontal;

— if double rungs, be fitted on the same horizontal level with a clear gap of not more than 50 mm between them.

— 3. The stringers or uprights of ladders should:

— be smoothly finished;

— be in one continuous length where possible; if a join is necessary and a fishplate is used, it should be fitted on the inside of the stringers;

— be adequately supported from the structure at suitable intervals;

— extend at least 1 m above the landing place (figure 15); if this is not practical, an adequate handhold should be provided above the platform;

Figure 15.   Vertical ladder through opening

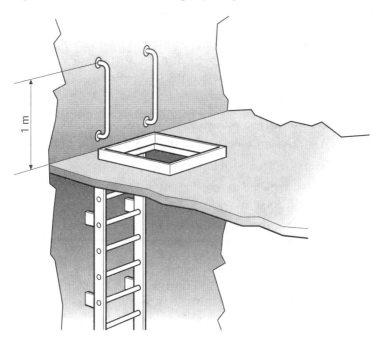

— be opened out above the platform to give a clear width of 700 to 750 mm to enable a person to pass through them (figure 16);

— be connected at their upper extremities to the guard rails of the landing platform or given other support as necessary;

— have a clear space of not less than 75 mm on each side for the user's hand.

Figure 16. Vertical ladder rising above edge

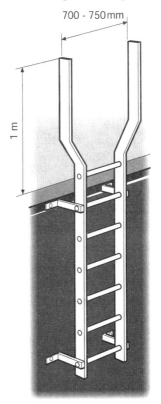

4. Where a landing platform is provided, it should:

— have a minimum dimension of 750 mm in either direction not less than 1 m above the floor of the platform enabling a person to stand safely on the platform;

— be protected on any open side by a rigid guard rail 1 m above the platform, an intermediate rail 500 mm above the platform and a toe board extending to a height of not less than 150 mm above the floor of the platform;

— have a floor with a non-slip surface;

— have a minimum headroom of 2.1 m.

5. A vertical ladder more than 3 m high should be fitted with guard hoops that should:

— be uniformly spaced not more than 900 mm apart;

— have a clearance of 750 mm from the rung to the back of the hoop;

— be connected by vertical strips secured to the inside of the hoops, the strips being equally spaced around the circumference of the hoop.

6. A vertical ladder more than 6 m high should be provided with suitable intermediate platforms at intervals of not more than 6 m.

7. A sloping ladder should not have a slope of more than 25° to the vertical.

8. A sloping ladder with a slope of more than 15° to the vertical should be provided with:

— treads or pairs of rungs. The front of a tread should overlap the next tread by at least 16 mm. Double rungs should be on the same horizontal level with a clear gap of not more than 50 mm between them;

— suitable handrails not less than 540 mm apart, measured horizontally;

— suitable guard hoops as for fixed ladders but with a clearance of 1 m from the front of the tread to the back of the hoop measured at right angles to the axis of the ladder.

9. All ladders and landing platforms should be adequately lighted whenever the ladder is in use.

10. Fixed walkways adjacent to roofs of fragile material should be securely fenced (see section 3.3.4).

11. Suitable warning notices prohibiting access to such roofs other than by crawling boards or other appropriate equipment should be displayed.

3.5.4. Portable ladders

1. Every portable ladder should be of sound material (rigid wood or metal, usually aluminium alloy), of good construction, of adequate strength, properly maintained and clearly identifiable, and inspected at suitable intervals by a responsible person. If dangerous defects are found, the ladder should immediately be taken out of service. Appropriate records should be kept of all inspections and repairs.

2. Every portable ladder (figure 17) should:

— have rungs equally spaced apart at intervals of not less than 250 mm or more than 350 mm;

— have rungs of a width between its uprights of not less than 380 mm or more than 450 mm;

— not exceed 6 m in a single length;

— not have more than two extending sections;

— if an extension ladder, be equipped with suitable guide brackets and an effective locking device so that each extension is securely held and locked in the desired position;

Figure 17.   Dimensions of portable ladders

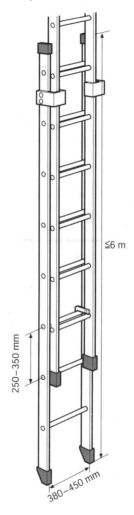

— if extended by ropes, have ropes that are securely anchored and run over pulleys having a groove suited to the rope size;

— not exceed 15 m in length when extended.

3. The rungs of a lightweight portable metal ladder should be:

— of adequate diameter (not less than 20 mm);

— secured to the uprights in such a way as to ensure that they do not turn or otherwise work loose;

— corrugated to minimize the danger of a person's foot slipping.

4. The uprights of a lightweight portable metal ladder should be:

— made in one continuous length;

— of sufficiently large cross-section to prevent dangerous deflection of the ladder in use;

— fitted with non-slip shoes or other suitable means of reducing to a minimum the likelihood of the ladder slipping.

5. A wooden ladder should:

— have uprights made of suitable wood with the grain running lengthways;

— have rungs properly secured to the uprights; nails or spikes should not be used;

— not be painted, but may be treated with a clear varnish or other effective preservative of a type that will not conceal any defect that would otherwise be visible;

— be provided with a sufficient number of metal cross-ties, where necessary.

3.5.5. Rope ladders

Rope ladders should not be used to provide access to places on shore.

3.5.6. Lifts

1. All lifts should comply with national legal requirements and be clearly marked with their maximum capacity. In passenger lifts this should be expressed in terms of both weight and the maximum number of persons.

2. At least one emergency stairway should be provided at each group of lifts.

3. Passenger lifts should be installed to provide access to the cabs of all new container quay cranes and similar large structures such as some large bulk-handling equipment. The installation should include provision for the rescue of any person in the lift in the event of a power failure or other emergency.

3.6. Terminal plant and equipment

3.6.1. General requirements

All terminal plant and equipment should be of good design and construction, of adequate strength, suitable for the purpose for which it is used, and maintained in a safe and efficient condition. The maintenance should be carried out on a planned preventive basis.

3.6.2. Mobile equipment

Mobile equipment used in ports, including various types of vehicles which are one of the most common elements in fatalities and serious injuries in ports, should be properly maintained and kept in good order. Special attention should be paid to the maintenance of brakes and braking systems.

3.6.2.1. Internal movement vehicles

1. Internal movement vehicles (vehicles that only work within a port or belong to the ship), including skeletal trailers, should comply with appropriate minimum standards for construction and maintenance, particularly with regard to such items as tyres, brakes, lights, steering, warning signals and general vehicle safety.

2. Cargo-handling vehicles should have a high degree of stability under working conditions.

3. Vehicles should be conspicuously painted or marked and fitted with a flashing or rotating yellow beacon.

4. Safe access should be provided to the driver's cab and to other parts of the vehicle as necessary.

5. Drivers' cabs on vehicles should provide protection from adverse weather conditions and have good all-round visibility, with minimal obstruction of the driver's view. Where vehicles have blind spots and there is a risk of injury, closed-circuit television or other suitable detection device should be considered.

6. All exposed dangerous parts of vehicles, such as power take-offs, chain drives and hot exhausts, should be securely guarded.

7. Vehicles, including trailers used to transport containers, should be constructed in such a way that the containers are supported by their corner fittings or other parts designed for that purpose. Containers should not be supported on their side rails.

8. Consideration should be given to the fitting of speed limiters to heavy-duty tractors and other plant for handling containers or similar large cargo.

3.6.2.2. Visual display screens in vehicles

1. Visual display screens in vehicles should be fitted in a location that is not prone to glare and reflection, and will provide the minimum distraction to the driver of the vehicle while it is moving but still be easily readable. The display should be kept concise and require as little interaction from the driver as possible when the vehicle is moving.

2. Any acknowledgement required from the driver should be simple and, where possible, interaction should only require the operation of a simple button, switch or touch screen.

3. Logistical systems should be designed so that any data input or complex keyboard operation by the driver can be carried out while the vehicle is stationary.

3.6.2.3. Skeletal trailers

1. Trailers used in port operations should be constructed so as to be able to withstand the continuous impacts when loaded by cranes or other container-handling equipment.

2. The safe working load (or maximum load rating) of trailers should be adequate for their use. Trailers used in container terminals may need to be constructed for a maximum load rating in excess of 50 tonnes.

3. Where trailers operate at night or in poorly lit areas, lighting and adequate reflectors for the trailers should be considered. Consideration should be given to light-emitting diode (LED) lighting systems. These have greatly improved

Figure 18. Container restraints on skeletal trailer

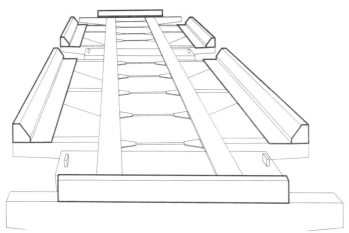

lighting efficiency and are not prone to loading impact damage.

4. All trailers should be fitted with devices to secure or retain loads on them.

5. Trailers should preferably be painted a conspicuous colour or otherwise be conspicuously marked.

6. The braking systems on trailers should be compatible with those on the tractors that are to move them.

7. Trailers that do not have conventional twistlock securing devices, and are used in container terminals where loads travel only short distances at slow speeds without negotiating sharp bends on roads, should be fitted with substantial corner plates or other restraints of sufficient height to retain the load in position (figure 18).

3.6.2.4. Trestles

1. Trestles (figure 19) should be used to support trailers that are not attached to other vehicles when the trailers are:

— laden;

— being loaded or unloaded by a lift truck from a loading bay;

— stowed on board ship.

2. The correct type of trestle with adequate strength for the task should be selected.

3. Trestles should preferably have wheels or rollers so that they can easily be moved. Wheels or rollers on heavy-duty trestles can be spring-loaded and not load-bearing when in use.

4. Trestles should be regularly inspected and maintained.

3.6.2.5. Goosenecks

1. All goosenecks should be regularly inspected and properly maintained, particular attention being given to the wear of the kingpin and the squared-off edge of the toe plate. A gooseneck with a missing or damaged toe plate should not be used.

2. Storage frames for goosenecks (figure 20) should be located so that tractor units do not have to emerge directly into a traffic stream when leaving the frame.

3. Storage frames for goosenecks may restrain them between vertical frames or by a shoe over the toe. Both are prone to damage and should be inspected and maintained accordingly.

Figure 19. Trestle

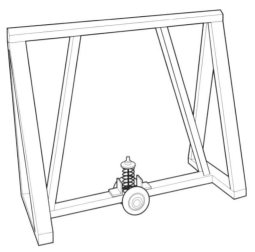

Figure 20. Storage frame for goosenecks

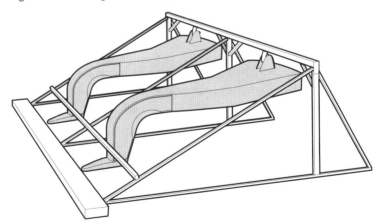

4. Many roll trailers have attachment arms suitable for laden and unladen conditions. Goosenecks may be modified or purchased with corresponding attachment lugs to ensure improvement of security, particularly when pulling up gradients.

3.6.2.6. Roll trailers and cassettes

1. Port operators should ensure that arrangements with roll trailer and cassette owners include procedures for their inspection and maintenance.

2. Roll trailers, and cassettes (used for forest products), should always be stored on firm and level ground.

3.6.2.7. Hand trucks and trolleys

1. Hand trucks used for transporting gas cylinders, carboys or similar objects should be designed and constructed for that specific purpose.

2. Hand trucks or trolleys used on slopes should be provided with effective brakes.

3. If it is advisable to prevent hand trucks and trolleys from moving when they are left standing, they should be provided with effective handbrakes, chains or other appropriate devices.

4. Three-wheeled or four-wheeled hand trucks should be provided with spring clips or other locking devices by which the handles can be secured in an upright position. Portworkers should be required to use these devices when the trucks are stationary.

5. Handles of hand trucks and trolleys should be designed to protect the hands of the user. This may be by the provision of knuckle-guards.

3.6.2.8. Cargo platforms

1. Cargo platforms should be robustly constructed of sound metal or wood. If the platforms are designed to be movable, perforated sheeting, expanded metal or metal grating should be used to minimize the weight of the platform. Platforms should be designed to bear the weight both of the loads to be made up or received and of the workers. A factor of safety should be allowed for the dynamic loads that will occur when cargo is landed on them.

2. Cargo platforms (figure 21) should be:
— adequately supported and, where necessary, securely fastened;
— of sufficient size to receive cargo and to ensure the safety of persons working on them;
— provided with safe means of access, such as ladders or steps;
— securely fenced (see section 3.3.4) on any side that is not being used for receiving or delivering cargo, if the platform is more than 1.5 m high;
— maintained in good repair.

3. Any portable trestles used to make up cargo platforms should be so placed as to be steady.

4. Cargo platforms should not be overloaded.

5. Hatch covers should not be used in the construction of cargo platforms.

Figure 21.   Cargo platform

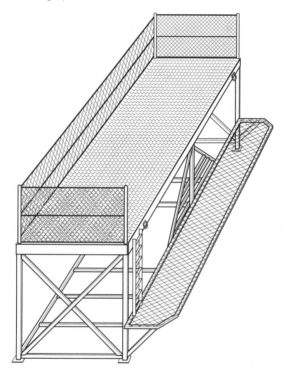

3.6.2.9. Access or lashing cages

1. The framework of most access or lashing cages (figure 22) is similar to that of an ISO container. The location of corner fittings in the top framework should conform to *ISO 668 Series 1 freight containers – Classification, dimensions and ratings* to enable the cage to be lifted by a container spreader.

Figure 22.   Access or lashing cage

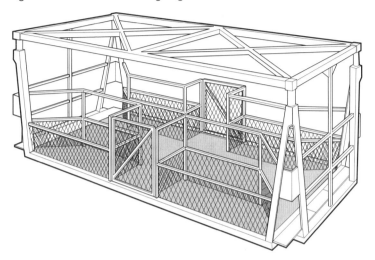

Most cages are 20 or 40 feet long but some telescopic cages have also been built. Smaller cages, sometimes known as gondolas, are used for work in narrow aisles between stacks of containers.

2.   Access or lashing cages should have:

— guard rails and toe boards (see section 3.3.4). The top rail should be recessed or an additional handrail provided inside the guard rail, in order to prevent hands being trapped between the guard rail and a container or other object. The distance between a handrail and guard rail should not be less than 90 mm in order to allow for workers wearing gloves;

— robust doors or gates that open inwards and are self-closing. Chains should not be used instead of doors;

— mesh or other suitable protection on the sides and ends of the cage to prevent accidental trapping;

— where practicable, protection from objects falling from above;

— suitable bins and hooks to stow equipment normally carried in the cage. Such equipment includes twistlocks and other inter-box connectors and twistlock poles. Covers for bins may form seats. The use of seats enables workers to be more stable during transfer to or from a quay;

— a secondary means of locking onto a spreader when the cage is in use (see section 7.8.3). The following possibilities may be considered:

• manual attachment of a chain at each corner;

• the use of hand-operated locking pins;

— provision of an additional automatically operated twist-lock at both ends of the spreader;

— a notice giving instructions for safe use.

3. The bottom corners of the cage may be recessed and the end of the cage floor may be hinged to allow safe access to twistlocks, etc., below the cage. Any area of floor where workers kneel should be suitable for the purpose and not of open-grid construction. A handrail should be provided in front of the kneeling position.

4. An emergency stop button should be provided, where possible.

5. Radio communication with the crane operator should be provided.

3.6.3. Conveyors

1. All dangerous parts of conveyor systems should be securely guarded. Enclosure of such parts also prevents the ingress of dust or other materials. The dangerous parts include:

— all drives;

— in-running nips between belts and end rollers or tension rollers;

— intakes between belts and other moving parts and stationary parts or other objects;

— intake openings of blowers or exhaust fans for pneumatic conveyors.

2. Enclosure of intermediate lengths of belt conveyors is not always necessary but may be needed to protect the material being conveyed from the weather or from being stripped off by wind.

3. Horizontal conveyors at floor level should be protected by metal grating or otherwise guarded.

4. Stopping devices should be provided at all workstations at power-driven conveyors. Trip wires should be provided along the side of the conveyor where walkways are alongside them. These should operate stop switches fitted at intervals of not more than 50 m.

5. The controls of any system of two or more conveyors operated together should ensure that no conveyor can feed onto a stationary conveyor.

6. Conveyor systems started remotely should have audible or visual warning systems to warn workers that the

system is about to start. Workers should be able to communicate with the control room. Closed-circuit video systems may be helpful.

7. Conveyor systems that can be started remotely or from more than one position should be fitted with lock-off switches in appropriate locations to protect persons cleaning or working on the system.

8. Provision should be made for the safe cleaning of conveyors and for clearing obstructions. Where appropriate, guards should be interlocked. A suitable time delay should be incorporated if the machinery has a significant overrun before stopping.

9. Walkways adjacent to open conveyors should be at least 1 m wide.

10. Suitably fenced bridges should be provided where it is necessary for workers to cross over conveyors.

11. Sheet or screen guards to catch any falling material should be provided at places where conveyors pass over workplaces or walkways.

12. Where the tops of hoppers feeding conveyors are less than 900 mm above the floor, the openings should be fenced.

13. The sides of conveyors moving solid objects should be at least 100 mm high, or half the height of the objects if greater.

14. Enclosed conveyors that convey flammable materials should be suitably explosion protected. This may be done by making the enclosure sufficiently strong to withstand and contain any explosion inside it, or by providing appropriate

explosion relief venting to a safe place, preferably in the open air. The enclosure should include facilities to enable any fire in it to be tackled.

3.6.4. Electrical equipment

1. All electrical equipment and circuits should be so designed, constructed, installed, protected and maintained as to prevent danger in accordance with national legal requirements. Where such requirements do not exist, reference should be made to the relevant recommendations of the International Electrotechnical Commission (IEC).

2. Suitably located efficient means for cutting off all electrical current from every part of the system should be provided as necessary to prevent danger.

3. All non-current-carrying metal parts of electrical equipment should be earthed, or other suitable measures taken to prevent them from becoming live.

4. All conductors should be suitably insulated and installed to prevent danger. Cables should be installed so as to be protected against being struck by moving loads.

5. Electrical equipment that will be exposed to wet or dusty conditions should be so constructed as to be suitable for use in such conditions.

6. Electrical equipment for use in places where a potentially explosive atmosphere is likely to occur should be constructed so as to be not liable to ignite that atmosphere. It may be possible to construct electronic and similar equipment so as to be intrinsically safe and unable to ignite the atmosphere. Other electrical equipment should be pressurized

or otherwise explosion-protected to a standard appropriate for use in that atmosphere.

7. Portable floodlights, and hand lamps in particular, should be powered as far as possible from a low-voltage circuit. The voltage should not exceed 42 volts between conductors, or 24 volts between phase and neutral in the case of three-phase circuits. In accordance with the recommendation of the IEC, the no-load voltage of the circuit should not exceed 50 volts in the first case and 29 in the second.

3.6.5. Hand tools

1. All manual and power-operated hand tools should be of good material and construction, and maintained in a safe condition.

2. Hand tools, including those owned by individual workers, should be periodically inspected by a competent person. Defective tools should be immediately replaced or repaired.

3.6.6. Machinery

1. All dangerous parts of machinery and pipes containing hot fluids, including those in places where only maintenance personnel work, should be securely guarded unless their position or construction makes them as safe as they would be if they were securely guarded. Dangerous parts of machinery include motors, gears, chains, wheels and shafts.

2. Guards should be of rigid metal or other material capable of withstanding the corrosive effects of the marine environment. Wood is liable to rot and is generally not suitable.

3. Guards that are not an integral part of a machine should be securely fastened in position to prevent unauthorized removal.

4. Machinery should be designed to come to a stop if a guard that is necessary is removed. Devices ensuring this should fail to safety. Limit switches should be positively operated so as to be driven to the open position.

5. Every machine should have a stop control next to the operator's position.

6. Pipes that are likely to reach temperatures above 50°C should be insulated.

3.6.7. Mooring dolphins and bollards

1. Safe access should be provided to all mooring dolphins and bollards.

2. A ladder from the water should be provided at all dolphins.

3. All dolphins and any walkways between them should be fenced (see section 3.3.4) as far as is practicable.

4. Appropriate life-saving equipment should be provided on or immediately adjacent to mooring dolphins.

5. Where appropriate, bollards should be clearly marked with an identification number. This should be clearly visible to mooring parties both on shore and on ships.

6. The ground around all bollards should be maintained in a sound and level condition.

7. The adequacy of available mooring facilities should be reviewed when changes are proposed to the type or size of ships to be handled in a port.

3.6.8. Vehicle sheeting facilities

1. Suitable vehicle sheeting facilities should be provided at ports where it is necessary to sheet or unsheet vehicles, and this cannot be carried out from the ground.

2. The simplest sheeting facilities consist of two platforms between which the vehicle can be driven (figure 23). The platforms should be of the same height as the bed of most freight vehicles. The outer edges of the platforms should be fenced (see section 3.3.4) and steps for access should be provided at each end.

3. Sheeting platforms may be built as permanent structures or designed to be dismantled and re-erected elsewhere.

3.6.9. Other equipment

1. All storage racks should be of good robust construction and of adequate strength. The racking should be cross-braced and firmly secured to the ground or other substantial structure to prevent collapse or overturning.

2. The racking should be clearly marked with its maximum safe load, where appropriate.

3. The construction of the rack and the weight and nature of the goods to be kept in it should be considered when determining the height/base ratio of racks.

4. Where mechanical handling equipment is used to load and unload racking, protection should be provided for exposed uprights of the racking in order to prevent damage to them that could result in the collapse of the racking. Protection of uprights at corners is particularly important.

Figure 23.   Vehicle sheeting platform

3.7. Bulk cargo terminals

3.7.1. Bulk solids

1. Special consideration should be paid to the stresses set up by solid bulk materials, and transmitted to the walls and foundations of structures. The walls of quays, bins and rooms where such materials are kept have all been known to collapse, particularly if heavy materials such as scrap are kept near the edge. The weight of cargo-handling vehicles that will be used in the area should be included in calculations.

2. Steps should be taken to prevent or minimize the escape of dust when solid bulk materials are being handled (see section 6.19 and Chapter 9). These may include enclosure of handling equipment such as conveyors, suction legs and elevators.

3. Plant handling solid bulk materials should be explosion-protected, as necessary. Buildings in which such plant is located should be kept clean at all times.

4. Pits from which elevator systems are fed and into which persons could fall should be securely fenced (see section 3.3.4).

5. Doors on upper access compartments of silos should be interlocked with the feed to stop the filling of the compartments as soon as an access door is opened.

3.7.2. Bulk liquids and gases

1. All bulk liquid and gas installations should be located and laid out in accordance with national legal requirements, and international and national industry standards and codes of practice. Particular attention should be given to the location and nature of neighbouring premises, the potential

effects of shipping that may pass the installation and the prevention of pollution from leakages or spillages.

2. An isolating valve should be installed in pipelines at the foot of each jetty and close to each cargo-loading arm. Where appropriate the valve should be able to be operated remotely in the event of an emergency (see section 6.11).

3. The safe operating envelope within which cargo-loading arms can be safely operated should be established.

4. All hoses should be tested and thoroughly examined periodically, in accordance with the manufacturer's and industry's recommendations.

5. All shore pipelines to which cargo hoses can be connected should be clearly identified (figure 24).

Figure 24.   Identification of shore pipelines

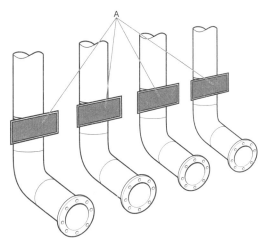

A. Identification plate

6. Electrical equipment at berths where bulk liquids are handled should be of appropriate explosion-protected construction for the cargoes to be handled.

7. An international ship–shore connection (see section 3.1.4.4, paragraph 9) enabling shore fire mains to be connected to the fire main on a ship should be provided at all tanker berths.

3.8. Container terminals

3.8.1. Definitions

1. The following definitions apply to container terminals and to the handling of containers in ports:

— *Avenue* is a marked access route within a container-stacking area for movement of transfer and stacking equipment between storage blocks. Also referred to as an "aisleway".

— *Block* is a rectangular marked and identified storage area within a stacking area for stacking designated groups of containers. A block is divided into rows by aisles.

— *Container-handling area* is the entire area in which containerized cargo is handled or stacked.

— *Control centre* is the administrative office from which port operational activities are controlled, usually by direct radio or computer communications.

— *Grid* is part of a container terminal to which access by road vehicles is permitted in order to deliver containers to, and collect containers from, container-stacking areas. Also referred to as an "interchange area".

— *Stacking area* is a storage area of a port behind a quay, in which containers are stacked to await onward movement. Also referred to as a "container yard".

— *Straddle carrier* is a tall, wheeled frame, wide enough to move astride a container and lift it by means of a suspended spreader; used for stacking and transferring containers.

— *Slot* is a clearly marked-out sub-area of a grid in a container-handling area, just sufficient in size to accommodate one road vehicle of maximum size; or an individual storage location on a container ship, e.g. a cell in a cell guide storage system, uniquely numbered for identification. The term is also used to identify a specific bay and row on a container ship.

3.8.2. General requirements

1. Container terminals should be laid out and organized in such a way as to separate persons on foot from vehicles, so far as is practicable.

2. Runways of rail-mounted or rubber-tyred gantry cranes should be clearly marked on the ground. It should not be possible to confuse the markings with those of a safe walkway.

3.8.3. Segregation

1. Except at the quayside, the operational areas of a container-handling area should be enclosed by a fence at least 2 m high, or other suitable means, to separate it from other activities in the port and prevent unauthorized entry. The fence should be of chain-link mesh or other suitably strong permanent construction.

2. Arrangements should be made to ensure that persons who need to enter the operational areas are able to do so safely. This may be achieved by the provision of clearly marked walkways that do not enter container-stacking areas or cross vehicular traffic routes, or by the provision of transport to their destination. Particular attention should be given to the need for access to ships at berths by ships' crews, mooring parties, pilots and other persons, and to blocks of refrigerated containers by refrigeration engineers.

3. Traffic routes in container-handling and container-stacking areas should be laid out to be one-way, as far as it is practical to do so. Traffic lights should be provided, where necessary.

4. If walkways necessarily cross traffic routes, appropriate markings and signs should be provided on the walkways and traffic routes to warn both pedestrians and drivers. Where traffic lights are provided, they should give precedence to vehicles.

5. The need for the vehicles of hauliers to enter container-stacking areas should be avoided as far as is practicable. This may be done by the provision of exchange grids where vehicles are loaded or unloaded, for example by straddle carriers.

6. Each container block and row should be clearly identified by markings on the ground or elsewhere. The markings should be maintained so as to be clearly visible to vehicle and crane operators at all times.

7. Obstructions in container-stacking and container-handling areas should be kept to a minimum. Any lamp standards or other obstructions that are necessary in such

areas should be protected by robust fencing that is clearly visible.

3.8.4. Reception facilities

1. Suitable facilities should be provided at the road entrances and exits to and from container-handling areas for checking of documents and the integrity of containers, including seals.

2. The building should preferably be designed so that the checker's window is at a convenient height for the drivers of container vehicles to permit the exchange of documents without the need for the driver to leave the cab of the vehicle.

3. The provision of suitable gantries at entrances and exits enables containers to be examined for security purposes and to be checked for twistlocks or other objects that have been left on the top of a container and could later fall off (figure 25). Mirrors and video cameras can also be used for these purposes.

4. Passengers in container vehicles should be prohibited from entering container-handling areas. A suitable room or area should be provided where passengers can await the return of the vehicle from the container-handling area.

5. A suitable area should be designated and clearly identified where vehicle twistlocks can be released in safety.

6. Suitable clearly signed and marked parking areas should be provided if vehicles can be expected to have to wait for significant times. If a parking area is situated at the side of a roadway, it should be sited so as to ensure that parked

Figure 25.   Gantry for checking containers

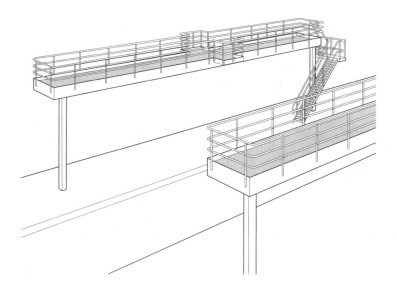

vehicles will not obstruct or restrict vision from vehicles on the adjacent roadway.

7. Clearly marked walkways should be provided from parking areas to welfare facilities or other areas or buildings which drivers may need to visit.

3.8.5. Control rooms

1. Control rooms for container-handling areas should have a good overall view of the area.

2. An efficient clear communication system should be provided between control and all terminal vehicles. In order

to minimize distractions to other drivers, the communication system should allow control to communicate with specific vehicles, rather than all vehicles all the time.

3.8.6. Grids

1. Suitable grids should be provided where straddle carriers are used to load or unload containers onto or from road vehicles.

2. The grids should be clearly marked and laid out in parallel or echelon formation with centres not less than 6 m apart.

3. Where practicable, the traffic routes for road vehicles and container-handling equipment should be laid out for a one-way traffic flow. Where this is not practicable and road vehicles have to reverse into a slot on a grid, the manoeuvring area should be sufficiently large to enable this to be carried out safely.

4. A safe area where drivers of road vehicles can stand while containers are being lifted onto or off their vehicles should be provided and clearly marked. The safe area should be located so that persons in it arc clcarly visible to drivers of straddle carriers as they approach it. If the area is located between slots, the size of the area should be determined in the light of the grid layout, the width of the straddle carriers and other relevant operational factors.

3.9. Passenger terminals

1. Special attention should be paid to the need to ensure the safety of passengers at cruise ship and ferry terminals.

2. Passenger access and exit routes should be clearly marked and laid out in accordance with national legal requirements. Internationally recognized visual symbols should be used.

3. Routes for foot passengers should be segregated from vehicle routes.

4. Areas to which the entry of passengers is prohibited should be clearly marked.

5. All public areas, walkways, ramps, lifts, bridges, etc., used by passengers should be clearly marked with any relevant weight or other limitations.

6. Access by passengers to the controls of ramps or other equipment should be prevented.

7. Appropriate facilities for the safe handling of passenger luggage should be provided. Where necessary, this should include facilities for security examination.

3.10. Roll-on-roll-off (ro-ro) terminals

1. Ro-ro terminals should be fenced off whenever practicable, with access controlled at suitable gates or barriers to prevent unauthorized access.

2. Ro-ro traffic should be controlled by road traffic signs, road markings and appropriate speed limits, as necessary. Speed limit signs should be repeated at appropriate intervals.

3. Traffic signs and road markings should conform to national road traffic requirements. Pictorial signs should be used whenever they are permitted, particularly in ro-ro term-

inals handling international services where the national language may not be the first language of many drivers.

4. Signs reminding drivers of the side of the road on which they should drive should be displayed at ro-ro terminals handling services from countries that drive on different sides of the road or that are near other countries that do so.

5. Suitable parking areas should be provided for vehicles waiting to board ships.

6. Trailer parks should be provided for unaccompanied ro-ro traffic. These should be separate from the holding or parking areas for accompanied ro-ro traffic.

7. Trailer parks should be laid out with a one-way traffic system, whenever practicable.

8. Pedestrian walkways between holding or parking areas and personal welfare facilities, and between coach parks and passenger terminals, should not cross road traffic routes. Any crossing that is necessary should be at right angles to the traffic route, and clearly marked and signed to warn both pedestrians and vehicle drivers. Traffic lights may be necessary at crossings from coach parks or other places that are frequently used by significant numbers of pedestrians.

9. At terminals where there are a number of ro-ro berths, each berth should be clearly numbered or otherwise identified.

10. The shore approaches to ramps of ro-ro ships should generally include a clear area of at least 35 m radius from the end of the ramp. This area should be clearly signed "Keep clear" and marked with cross-hatched yellow lines or

otherwise in accordance with national requirements. Where this is not practicable, temporary signs should be used.

3.11. Warehouses and transit sheds

1. All areas and buildings where goods are kept should be designed and constructed in accordance with national legal requirements. These should take into account the possible need for water sprinklers in a building and any regional climatic factors, such as the need for additional ventilation or insulation, the need to allow for snow loadings on roofs, etc.

2. All floors should be designed with adequate strength to support the maximum load of the goods and the handling equipment that are to be used on them. The design should take into account dynamic forces that may result from the landing of loads and the operation of handling equipment, as well as the static forces. If the maximum permitted loading is less than that generally permitted in the port areas, the limit should be clearly shown on prominent notices.

3. Balconies on which goods are landed on the outside of multi-storey buildings should be about 2 m wide. If there are separate balconies at individual openings, each should be at least 4 m long and 1 m longer than the width of the opening.

4. Storage areas should be laid out with suitable traffic lanes. These should have adequate clearances to permit the safe use of the relevant handling equipment.

5. Aisles should be clearly marked off with continuous yellow lines.

6. Where vehicles are driven into warehouses or sheds, there should be adequate clearance between the vehicle and any goods it may be carrying and the sides and tops of the entrances. If there is only limited clearance, the edges and top of the openings should be clearly highlighted by alternate black and yellow diagonal stripes or otherwise. Top markings are particularly important if lift trucks are used. There have been a number of fatalities at entrances with limited clearance when the top of the entrance has been struck by the mast of a lift truck being driven with the mast partially raised. If the width of the entrance is restricted, pedestrians and vehicles should be separated by the provision of a separate door for pedestrians. The routes for pedestrians and vehicles should be indicated by appropriate signs.

7. Wherever practicable, stairs and lifts in warehouses should be located alongside a wall in order to minimize obstruction of storage space and access routes for cargo-handling equipment.

8. All openings in floors and walls should be securely fenced (see section 3.3.4).

9. The covers of openings in floors should be robustly constructed and of adequate strength to support any persons or plant that may pass over them when they are closed. Loose sheets should never be used to cover openings in floors.

10. Handholds should be provided on each side of openings in walls, floors or roofs where fencing may need to be temporarily opened to permit the passage of goods.

11. Appropriate ventilation should be provided in buildings where vehicles with internal combustion engines are used or toxic, explosive or flammable goods are kept.

12. Bare crane conductor wires should not be installed in warehouses and transit sheds. A variety of fully insulated power supply systems are now available. Existing bare crane conductor wires should be guarded or the height of stacks of goods limited to prevent danger. If access near to bare conductors is necessary, the conductors must be isolated.

13. All doors of refrigerated chambers in warehouses should be able to be opened from the inside at all times. A bell or other suitable means for summoning help in an emergency should be provided, where necessary.

14. In countries where there is a legal requirement to maintain the low temperature of some foodstuffs throughout the transport chain, it may be necessary to construct special intermediate doors at warehouse loading bays to maintain the low temperature during loading or unloading.

3.12. Gatehouses and quay offices

3.12.1. Gatehouses

1. Gatehouses should comply with the national legal requirements for offices, particularly those relating to overcrowding, cleanliness, lighting and ventilation.

2. Particular attention should be paid to the ventilation of gatehouses. Of necessity, they are usually located on major traffic routes and so exposed to considerable quantities of vehicle exhaust fumes. Consideration should be given to instal-

ling positive pressure ventilation systems. These should take in clean air from an appropriate location and discharge it through the gatekeepers' grilles adjacent to the traffic.

3. The building should be constructed in such a way that the grilles through which gatekeepers need to speak or pass documents to visitors are at a convenient height for both the gatekeeper and the visitor.

3.12.2. Quay offices

1. Quay offices should comply with the national legal requirements for offices.

2. Particular attention should be paid to the location of quay offices, particularly that of small temporary offices.

3. Quay offices should not be located where they are likely to be struck by passing cargo-handling equipment or by falling containers in the event of high winds.

4. Safe walking routes should be provided to all quay offices. Where appropriate, these should be clearly marked.

3.13. Port railways

1. Port railways should be constructed, equipped and operated so as to be compatible with relevant national legal requirements governing railways, where appropriate.

2. Specialized training of railway workers should be provided in accordance with national railway legal requirements, when appropriate.

3. Ground levers working points should be so placed that persons working them are well clear of adjacent lines

and the levers cause as little obstruction as possible to any person who may be put at risk.

4. Point rods and signal wires should be covered or otherwise guarded, where this is necessary to prevent danger.

5. Warning signs and fencing around obstructions on port railways should be made conspicuous by painting or otherwise.

6. All warning signs and fencing around obstructions should be appropriately illuminated when port railways are operated during hours of darkness.

7. Fouling points beyond which vehicles should not be parked should be clearly indicated. They should be positioned at points where there is sufficient space for a person to pass safely between vehicles on one line of rails and vehicles on a converging line. Small markers at ground level are often used for this purpose.

8. Areas where railway wagons are parked should be as nearly level as practicable; any gradient on such lines should not exceed 0.5 per cent (1 in 200). If the line is a dead-end siding, any gradient should be down towards the buffers or end stops.

9. Working areas should, where applicable comply with the provisions of sections 3.1.2 (Surfaces) and 3.1.3 (Lighting), and should be maintained in a sound condition, especially where rails run alongside the quay. Where practicable, the surface should be level with the tops of the rails.

10. Adequate clear space should be allowed between railway tracks and structures, piles of cargo or material traffic routes and walkways. This should allow for the width of railway wagons and should not be less than 2 m.

11. Workers should be protected against stepping onto the rail tracks in front of moving vehicles by suitable barriers and warning signs where buildings have exits opening directly onto port railway tracks, at blind corners and at other places where the field of vision is particularly restricted. Where practicable, this should be achieved by fixed railings across the direct route (figure 26). There should be a safe clearance between the railings and shunters or railway workers riding on rail vehicles.

12. Loading platform edges should be painted in a light-coloured paint to highlight the drop in level or gap between it and a wagon, and the danger to both workers on foot and those operating vehicles on the platform. The paint should preferably be reflective.

13. Bridge plates spanning gaps between loading platforms and the floors of rail wagons (figure 27) should:

— be clearly marked with the maximum safe load to be carried;
— include positive stops or hooks to prevent slipping or other unintended movement;
— have non-slip surfaces;
— have toe boards at least 150 mm high at the sides;
— have handholds or other appropriate devices by which they can be moved or lifted.

Figure 26.   Protection of an exit from a building adjacent to a railway

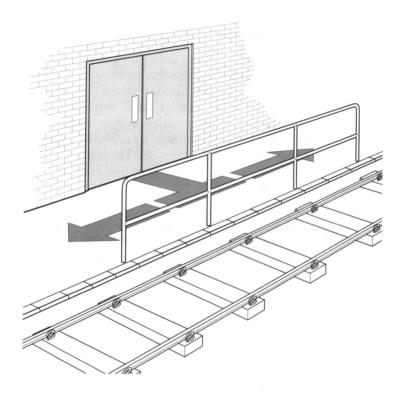

14. Where necessary, an appropriate loading gauge should be provided before the exit from a port railway to the national railway system to detect oversize loads. For instance, in some countries 9 ft 6 in containers can be carried on only a few rail routes.

Figure 27.   Bridge plate to rail wagon

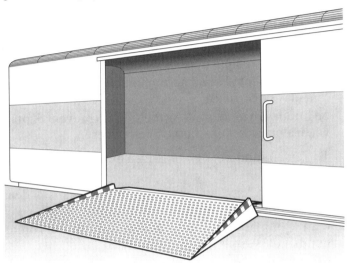

3.14. Tenders and work boats

1. All tenders, work boats and other craft used for the transport of portworkers should be of suitable construction, properly equipped for use and navigation and maintained in a seaworthy condition.

2. All tenders and work boats should comply with relevant national legal requirements. These generally relate to their construction, stability, life-saving equipment and operation.

3. All tenders and work boats should:

— be under the charge of a competent person;

— be manned by an adequate and experienced crew;

— display in a conspicuous place the maximum number of persons that may be carried; this should have been certified by a competent person;

— have sufficient seating accommodation for all persons that may be carried;

— be fitted with fenders along the sides to prevent damage when alongside ships, buoys or elsewhere.

4. Tenders and work boats driven by mechanical power should have:

— bulwarks at least 600 mm high or rigid rails at least 750 mm high to prevent persons falling overboard;

— seating under cover for at least half the maximum number of persons that may be carried;

— protection from the weather for the remainder of the persons wherever practicable;

— an appropriate number of suitable fire extinguishers;

— appropriate life-saving equipment.

3.15. Personal protective equipment

3.15.1. General requirements

1. Personal protective equipment (PPE) should never be used as a substitute for eliminating or otherwise controlling a hazard. However, when this is not possible, PPE should always be used.

2. The need for PPE should be determined in accordance with national legal requirements and an assessment of

the risks to which portworkers may be exposed during their work.

3. Cargo handlers should generally be supplied with:
— protective footwear;
— safety helmets;
— overalls;
— appropriate foul-weather clothing;
— high-visibility outer garments;
— gloves.

4. PPE that should be supplied to portworkers may include:
— ear/hearing protection;
— fall prevention and fall arrest equipment;
— flotation aids;
— foot and ankle protection;
— hand and arm protection;
— head protection;
— high-visibility clothing;
— knee and leg protection;
— overalls;
— respiratory protection;
— restraints;
— shoulder protection;
— weather- and heat-resistant clothing.

5. All PPE should comply with relevant international and national legal requirements and standards.

6. When selecting the most appropriate type of PPE for an operation, information available from equipment manufacturers and suppliers and those who will have to wear it should be considered before a final selection is made.

7. PPE should be selected to ensure that it is as comfortable as is practicable for those who have to wear it, as this can be onerous, particularly if more than one type of protection needs to be worn at the same time.

8. It is essential that appropriate training in the use, care and maintenance of PPE is given to all portworkers.

3.15.2. Storage and maintenance of personal protective equipment

1. Suitable facilities should be provided for the storage of PPE when it is not in use. This should generally be separate from accommodation for workers' personal clothing in order to prevent cross-contamination. The facilities should include means of drying, where necessary.

2. All PPE should be maintained in a clean, hygienic and effective condition, in accordance with the manufacturer's recommendations.

3. Some PPE has a limited useful life. Limitations may be based on time or the use of the equipment. In such cases, the equipment, or the relevant part of it such as the filter in a respirator, should be changed in accordance with the manufacturer's recommendations.

4. Lifting appliances and loose gear

4.1. Basic requirements

4.1.1. General requirements

1. Every lifting appliance and item of loose gear should be:

— of good design and construction, of adequate strength for its intended use and free from any patent defect;

— made to a recognized international or national standard;

— tested, thoroughly examined, marked and inspected in accordance with section 4.2;

— maintained in good working order.

2. Occupational safety is affected not only by the design of lifting appliances but also by that of their accessories and other loose gear used with them. The proper design and maintenance of all of them are essential, since breakage of any of them may cause serious accidents. Deterioration may be visible, as when it starts from the surface, or concealed internally; in either case, the mechanical strength of the material is reduced.

3. Documentation (as appropriate) relating to lifting appliances should include:

— a driver's instruction manual;

— an erection manual;

— a maintenance manual;

— a spare parts manual;

— the manufacturer's certification of fitness for use;

— a certificate of test and thorough examination after initial erection;

113

— the manufacturer's certificates for wire ropes installed on cranes;

— examination and maintenance records.

4.1.2. Brakes

1. Every power-operated lifting appliance should be provided with an efficient brake or brakes capable of stopping a load while it is being lowered.

2. The brakes should normally be applied automatically when:

— the motion control lever is returned to its neutral position;

— any emergency stop is operated;

— there is any power supply failure;

— in the case of electrically operated brakes, there is a failure of one phase or a significant drop in voltage or frequency of the power supply.

3. Band brakes generally act in a preferential direction and are sometimes jerky. They should only be used for emergency braking. Brakes with symmetrical jaws and two pairs of pivots have a gradual action.

4. A slewing brake should be capable of holding the jib stationary with the maximum safe working load suspended at its maximum radius when the maximum in-service wind acts in the most adverse direction. Sudden application of the brake should not damage the jib.

5. The brake lining or pads should remain adequately secured during their working life. Unless the brake is self-adjusting, appropriate means should be provided to permit brake adjustment to be readily carried out in safety.

6. The design of electrically operated brakes should ensure that the operating solenoid cannot be accidentally energized by the back electromotive force of any motor driven by the crane, by a stray or rogue current or by breakdown of any insulation.

4.1.3. Electrical supply

1. Self-reeling flexible cables should not allow long lengths of cable to drag on the ground where they can be exposed to damage. Outlets should generally be not more than 50 m apart. The use of motorized reels is preferable to springs or counterweights. Reels on quay cranes should be placed on the waterside, preferably on the outside of the gantry legs.

2. Trolley systems should be fed by overhead conductors or conductors in channels.

3. Overhead conductors should be sufficiently high to prevent contact by a vehicle or its load. Supports should be protected by suitable barriers where necessary.

4. Channels for conductors should be properly drained and designed to prevent entry of any object likely to cause danger.

4.1.4. Safe working load (SWL)

1. The safe working load (SWL) of all lifting appliances and items of loose gear should be based on the factors of safety set out in Appendix E.

2. Every lifting appliance and item of loose gear should be marked with its safe working load. The markings should be in kilograms if the safe working load is 1 tonne or less, or in tonnes if it is more than 1 tonne.

Figure 28.   Examples of marking SWL on heavy items of loose gear

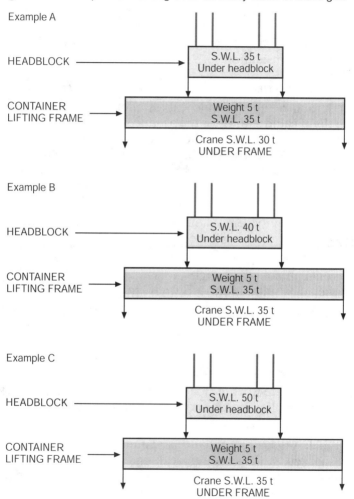

Example A

HEADBLOCK

S.W.L. 35 t
Under headblock

CONTAINER
LIFTING FRAME

Weight 5 t
S.W.L. 35 t

Crane S.W.L. 30 t
UNDER FRAME

Example B

HEADBLOCK

S.W.L. 40 t
Under headblock

CONTAINER
LIFTING FRAME

Weight 5 t
S.W.L. 35 t

Crane S.W.L. 35 t
UNDER FRAME

Example C

HEADBLOCK

S.W.L. 50 t
Under headblock

CONTAINER
LIFTING FRAME

Weight 5 t
S.W.L. 35 t

Crane S.W.L. 35 t
UNDER FRAME

3. Lifting appliances where the safe working load varies with the radius of operation should display a chart, showing the radius and the corresponding safe working load, in the cab in a position where the operator at the controls can clearly see it. The chart should also state the maximum and minimum operating radius for the appliance and from where the radius is measured.

4. Such appliances should also be fitted with a radius indicator that can be clearly seen by the operator at the controls and, where practicable, a safe working load indicator.

5. The maximum load that may be lifted when items of loose gear that have a significant weight (see section 4.2.6, paragraph 11) are attached to lifting appliances should be unambiguous (figure 28). There should be no confusion between the safe working load:

— below the header block/hook of the lifting appliance;

— of the loose gear;

— below the loose gear.

4.1.5. Controls

1. Controls of lifting appliances should conform to ISO 7752 *Lifting appliances – Controls – Layout and characteristics* and ensure that the operator has ample room for operation when at the controls.

2. Controls (figure 29) should be:

— so positioned that the operator has an unrestricted view of the operation or any person authorized to give the operator signals;

— marked with their purpose and method of operation.

117

Figure 29.   Example of lifting appliance controls – Ships' derricks

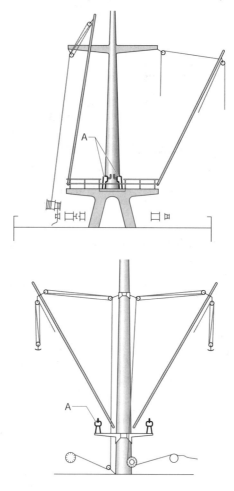

A. Controls.

3. The operating pedals for travel motions of mobile lifting appliances should follow road traffic practice with clutch (when fitted) on the left of the operator's feet, accelerator or other power control on the right and a brake between the other two pedals.

4. Whenever driving considerations permit, controls should return to the neutral position when released.

5. Consideration should be given to fitting "dead man's" controls to prevent inadvertent movement.

6. The control system should be such that no motion can start when the power supply is connected or the engine started. Movements should only be possible after a positive action.

4.1.6. Limiting devices

1. Limiting and indicating devices should conform to ISO 10245 *Cranes – Limiting and indicating devices*.

2. Wherever possible, every limiter should be positively actuated and designed to fail safe.

3. Where one motion of an appliance can cause a second motion to approach a limiter (e.g. a derricking-out motion that can cause a hoist motion to reach its limit), the limiter should stop both motions.

4. Every power-operated lifting appliance other than a ship's derrick should, where practicable, be fitted with a safe working load limiter. This should operate when the load being raised or lowered exceeds the safe working load by a predetermined amount, generally within the range of 3 to 10 per cent above the safe working load. The limiter should only prevent motions that would increase the overload.

5. Cranes should also be fitted with the following limiters:

— *hoisting limiter* preventing the load-lifting attachment being raised to the position where it strikes the structure of the crane;

— *lowering limiter* ensuring that the minimum number of turns is always left on the winch drum;

— *derricking-in limiter* ensuring that the crane jib cannot be derricked back beyond the minimum radius position;

— *derricking-out limiter* ensuring that the jib cannot be derricked out beyond the maximum radius position;

— *trolley or crab limiter* ensuring that the trolley or crab is stopped before it reaches the track end stops;

— *slewing limiter* on cranes with a limited arc of slew;

— *long travel limiter* on rail-mounted cranes preventing them from approaching the track end stops.

4.1.7. Lubrication

Every greasing and lubrication point should be located where lubrication can be carried out safely. Remote lubrication points should be provided where necessary.

4.1.8. Operator's cab

1. The operator's cab should provide the operator with a safe and comfortable working environment. The cab and its fittings should be constructed of fire-resistant material and conform to ISO 8566 *Cranes – Cabins*. In particular it should have:

— an unrestricted view of the area of operation;

— adequate protection from the elements;

— windows that can be readily and safely cleaned inside and out;

— a windscreen wiper on any window that normally affords the operator a view of the load;

— a comfortable seat that enables the operator to look in the required direction;

— a sliding or inward-opening door readily openable from inside and outside if the cabin is elevated;

— means of emergency escape;

— suitable fire extinguishers.

2. The operator's cab should be designed to limit noise and vibration to within nationally permitted levels.

4.1.9. Overhauling weight

An overhauling weight fitted at the end of a hoist rope should be:

— designed to minimize the danger of it catching on any part of a hold, ship's superstructure or similar obstruction;

— connected to the hoist rope by means of a short length of chain, where practicable.

4.1.10. Swivels

1. A swivel should be fitted between the hoist rope and the lifting attachment.

2. The swivel should be fitted with ball bearings or roller bearings that can be regularly lubricated (figure 30).

Figure 30.   Swivel

4.1.11. Tyres

1. Tyres of lifting appliances that travel on wheels should be selected to be appropriate for the intended duty.

2. Radial and cross-ply pneumatic tyres should never be mixed.

3. The correct pressure of pneumatic tyres should be conspicuously marked near each wheel.

4.1.12. Access

Safe means of access to all working positions on lifting appliances should be provided.

4.1.13. Winch and rope drums, leads and anchorages

1. Winch drums should conform to ISO 8087 *Mobile cranes – Drum and sheave sizes*.

2. Ropes should be fastened to winch drums in the manner prescribed by their makers.

3. The derricking and hoisting drums of a ship's derrick or derrick crane should be capable of accommodating the maximum working length of rope and the number of complete turns to remain on the winch specified by the manufacturer.

4. The angle of a wire rope lead to a winch drum should be sufficiently small to ensure that the rope is not damaged in service. The angle between the rope and the plane perpendicular to the axis of the drum should generally not exceed 1 in 16 for hoisting ropes and 1 in 12 for derricking ropes.

5. Where it would otherwise not be possible to avoid an excessive lead angle, a suitable coiling or spooling device should be fitted.

6. Lowering operations should normally be possible only with the winch connected to the power source. Free-fall lowering should be possible only in exceptional circumstances and if the winch is equipped with an automatic speed-limiting device.

4.1.14. Maintenance

1. All lifting appliances and loose gear should be maintained in good working order, and in efficient condition and good repair.

2. Maintenance, including lubrication, should be carried out on a regular scheduled basis, in accordance with the manufacturer's recommendations and operational experience.

3. Replacement components should conform to the manufacturer's manual or be of an equivalent standard.

4. Repairs to the structure of a lifting appliance should follow the correct procedure specified by the manufacturer. Excessive heat can change the properties of steel.

5. Corrective maintenance should also be carried out when necessary.

6. An accurate record of all routine and corrective maintenance should be kept.

4.2. Testing, thorough examination, marking and inspection of lifting appliances and loose gear

4.2.1. Introduction

Lifting appliances and loose gear should be safe when first provided and should remain safe throughout their operational life. The procedures for achieving this are well established, based on testing, thorough examination, marking and inspection. It is widely accepted that the testing of certain types of loose gear should be subject to a different testing regime. The present requirements and current good practice should be seen as core requirements for safe operations in port work.

4.2.2. Testing of lifting appliances

1. All lifting appliances should be tested in accordance with Appendix A and national requirements before being taken into use and after any substantial repair.

2. Lifting appliances should be re-tested:

— at least once in every five years, if part of a ship's equipment;

— as prescribed by the competent authority, if shore based.

3. The testing of cranes should be carried out in accordance with *ISO 4310 Cranes – Test code and procedures*.

4. The tests should cover all parts, and should be supplemented with a detailed examination of the appliance as a whole. The tests are matters for specialists and should be carried out by organizations whose competence has been recognized.

5. All assembled parts of a lifting appliance should be tested under a proof load, in accordance with Appendix D, section D.1.

6. The test conditions for the various parts should be those imposing the severest stresses on each part when in service. Derricks should be tested at the lowest practicable angle to the horizontal, and the slewing motion of an appliance with a derricking jib should be test braked at the lowest practicable angle of the jib.

7. A record of all tests of lifting appliances and related certificates should be kept and should be available.

8. The content and layout of the documents should be as established by the competent authority and in accordance with the model documents recommended by the ILO.

9. All loose gear attached to a lifting appliance should be tested in accordance with section 4.2.3.

4.2.3. Testing of loose gear

1. All loose gear should be tested in accordance with Appendix B and national requirements before being taken into use and after any substantial alteration or repair.

2. Wire ropes taken into use should:

— be made to a recognized national or international standard;

— have their minimum breaking load certified by the maker;

— be of a construction suitable for the purpose for which they are intended.

3. Requirements for the testing of wire ropes are generally set out in national or international standards. Wire rope slings with hand-spliced or mechanically secured eyes (with aluminium or steel ferrules) should be made from wire, manufactured to a recognized national or international standard, and supplied with a manufacturer's certificate showing the minimum breaking load before the termination or eyes were made. Wire rope slings with ferrule-secured eyes should be subjected to a proof test not exceeding twice the rated safe working load in straight pull.

4. Hand-spliced wire, fibre rope and webbing slings should be made from wire or fibres manufactured to a recognized national or international standard and supplied with a manufacturer's certificate showing the minimum breaking load. These slings are not supplied with a manufacturer's test certificate. They should never be subjected to a proof load that exceeds their safe working load. Proof loads in excess of the safe working load are carried out on slings made from synthetic fibres by the manufacturer on a batch basis.

5. Wire rope slings with ferrule-secured eyes should be individually tested.

4.2.4. Thorough examination

1. Thorough examinations of every lifting appliance and item of loose gear should be carried out periodically by a competent person, in accordance with Appendix C.

2. Lifting appliances should be thoroughly examined at least once every 12 months or after any repair or modification. Appliances used to lift persons should be thoroughly examined at least every six months, or at shorter intervals determined by a competent person.

3. Loose gear should be examined at least once every 12 months, or at such shorter intervals as may be prescribed by the competent authority or competent person, and after any repairs or modifications. These examinations should include hammer tests, removal of paint to expose the metal underneath, ultrasonic examination, radiographic examination and the dismantling of concealed components where appropriate.

4. Wire ropes and chains should be examined more frequently. Some users use ropes rather than chains, since ropes show up wear more easily and broken wires project from the rope.

5. If a wire rope contains any grips, wedge sockets or the like, they should be removed during the examination of the rope.

6. The thorough examination of blocks will usually require the block to be stripped and the pin examined.

7. Periodic examinations afford opportunities for deciding whether chains should be discarded or, in the case of wrought iron, sent for heat treatment.

4.2.5. Test and examination reports, registers and certificates

1. The results of tests and examinations should be recorded.

2. After completion of the thorough examination, the competent person should prepare a report which:

— clearly identifies the item examined, the date of the thorough examination, its safe working load(s) and any defects found;

— specifies any parts to be repaired or replaced;

— includes a statement that the item is, or is not, safe for continued use;

— gives the date by which the next test and thorough examination of a ship's lifting appliance should be carried out;

— gives the date by which the next thorough examination of all other lifting appliances and loose gear should be carried out;

— gives the name and qualifications of the competent person;

— includes any additional items required by national legislation.

3. The model form of register for ships' lifting appliances and certificates as required by Article 25 (2) of Convention No. 152 replaced earlier versions in 1985. Competent authorities in many countries have issued their own registers, complying with the ILO model form by giving the English text side by side with a translation into their own language.

4. Such records only provide evidence of the safe condition of lifting appliances and loose gear at the time of the examination.

5. Records should be kept on shore or on ship, as appropriate.

6. Registers and certificates for gear currently on board ship or on shore should be kept for at least five years after the date of the last entry.

7. Records may be kept in electronic form, provided that the system includes a means of making them available and of verifying the record.

4.2.6. Marking

1. All lifting appliances should be legibly and durably marked with their safe working load.

2. On derricks, the lifting capacity should be shown near the seating (gudgeon pin) in painted letters and figures within a frame of indentations or welding spots incised on a brass plate or inscribed on other material sufficiently resistant to defacement. On cranes, the capacity should be painted on metal plates that are then enamelled or covered with varnish.

3. Every item of loose gear should be legibly and durably marked in a conspicuous place with its safe working load, with an alphanumeric identification mark to relate it to records of test examinations and, where appropriate, with a mark to indicate the quality grade of the steel from which it is made. Where appropriate, the inscriptions should be incised, stamped or outline-welded.

4. The marking should be made in a place where it will not give rise to stress.

5. On long chains, the markings should be in a number of places.

6. The quality grade mark on steel items should be in accordance with Appendix F.

7. Where the markings are stamped directly on the gear, the stamps should not exceed the following dimensions:

Safe working load of gear	Maximum size of stamp (mm)
Up to and including 2 tonnes	3
Over 2 tonnes and up to and including 8 tonnes	4.5
Over 8 tonnes	6

8. Where stamps are used on chain links, the stamp size should not exceed the following dimensions:

Link diameter	Maximum size of stamp (mm)
Up to 12.5 mm	3
Over 12.5 mm and up to 26 mm	4.5
Over 26 mm	6

9. The stamp should give a concave indentation without sharp corners, and should not be struck with a blow greater than is necessary for a clear indentation.

10. If the material is too hard or if direct marking would affect or be liable to affect the subsequent safe use of the gear, the marking should be made on some other suitable item of durable material permanently attached to the gear,

such as a tablet, disc or ferrule. Marking on such items may be larger than the dimensions indicated in paragraphs 7 and 8 above.

11. Larger items, such as lifting beams, container spreaders or similar gear, that have a significant weight should also be conspicuously marked with their own weight. The markings should be so positioned and of such size as to be immediately legible to those using the gear from the quay or ship's deck.

12. Wire ropes used in long lengths without terminations are not usually marked. The manufacture's certificate for the wire is endorsed with its place of use to enable identification. A wire or wire sling with a thimble or loop splice ferrule should be proof-loaded and the safe working load stamped on the ferrule.

13. Markings on slings should be made in a permanent manner on:
— the terminal ring or link;
— a tablet, disc or ferrule attached to the sling, provided that the attachment will not cause damage to the rope;
— a ferrule of a wire rope having ferrule-secured eyes;
— the sling itself;
— a label; or
— by an approved electronic capture system.

14. Markings on slings should include the number of legs and the safe working load in straight lift and when the angle between the legs and the vertical is 45°.

15. Non-metallic slings should be marked with a label. The label should show, or have electronically captured, the sling's:

— safe working load in straight lift;
— material;
— nominal length;
— individual identification mark and traceability reference;
— manufacturer's or supplier's name.

16. Single-sheave blocks should be marked in accordance with Appendix H.

4.2.7. Inspection

All lifting appliances and loose gear in use should be regularly inspected by responsible persons (see sections 5.1.4.2 and 5.1.4.3). The inspections should comprise visual examinations to check that, so far as can be seen, the equipment is safe for continued use.

4.3. Lifting appliances

4.3.1. Ships' lifting appliances

1. Every ship should carry adequate rigging plans showing at least:
— correct position of guys;
— resultant force on blocks and guys;
— position of blocks;
— identification markings of blocks;
— arrangements for union purchase (where relevant).

2. Safe operation of ships' derricks (figures 31-34) depends largely on the proper maintenance of the running rigging. Wear and tear should be reduced as far as practicable. It is essential to ensure that running ropes do not rub against a fixed or mobile part.

Figure 31.   Two types of ships' derricks

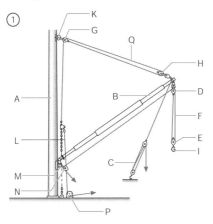

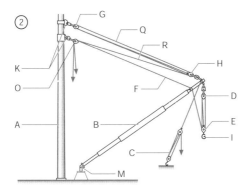

1. Light type.   2. Heavy type.

A. Mast or post. B. Boom. C. Pendant. D. Upper hoisting pulley. E. Lower hoisting pulley. F. Hoisting rope. G. Upper pulley. H. Lower pulley. I. Hoisting hook. K. Swivel for upper pulley. L. Pendant for topping lift. M. Bottom swivel. N. Lead block (guide pulley) on derrick heel. O. Return pulley. P. Heel block. Q. Topping lift. R. Topping lift preventer.

133

Figure 32. Ships' lifting appliances – A ship's derrick with topping lift and pendant

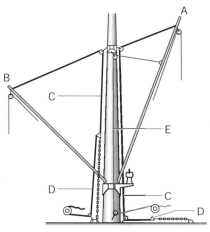

A. Long boom. B. Short boom. C. Topping lift. D. Pendant. E. Auxiliary lift (to move boom to and from working position).

3. Heel blocks should be restrained by a tensioning device to prevent them from swinging down during lowering when there is no load on the rope.

4. A derrick should neither be rigged nor used at an angle less than the minimum angle marked on it.

5. Derricks should be rigged in such a way that their components cannot whip against the winchman.

6. It should be ensured that light derrick booms do not lift out of their seating.

7. Each derrick should be legibly marked with its safe working load, as follows:

— used only in single purchase SWL xt
— used additionally with a lower cargo block SWL x/xt
— used in union purchase SWL (U) xt
(where x = safe working load).

8. The lowest angle to the horizontal at which the derrick may be used should also be marked on the derrick.

9. The letters and numbers should not be less than 770 mm high and should be painted in a light colour on a dark background, or in a dark colour on a light background.

10. The winch operator's stand should be protected against the weather by a cab with large windows.

11. The derrick luffing winch should have an effective blocking arrangement. This normally consists of the traditional pawl engaging in the wheel. Whatever device is used, it should eliminate all risk of loss of control during the raising or lowering of a load.

12. A ship's cargo lift should have controls:
— of the "dead man's" type that fail safe;
— arranged so that only one set of controls can be operated at a time;
— placed so that the operator is:
 • not in danger from the lift or moving vehicles; and
 • able to see the whole of the lift platform at all times.

13. An independent emergency stop control should be fitted in a prominent position among or near the other controls.

14. Each opening in a deck for a cargo lift should be protected by barriers that are:

Figure 33.   Rigging of ship's derrick boom

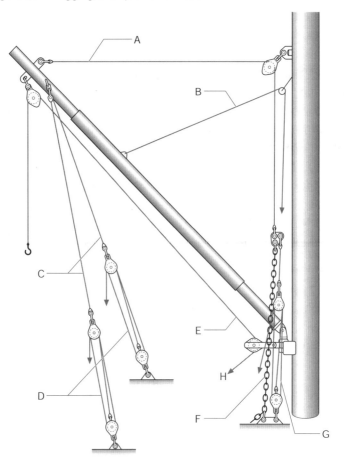

A. Topping lift. B. Auxiliary topping lift used to move the boom to and from its working and stowed positions. C. Pendants. D. Pendant blocks. E. Cargo runner. F. Span chain. G. Topping lift block. H. To winch.

Figure 34. Samson post

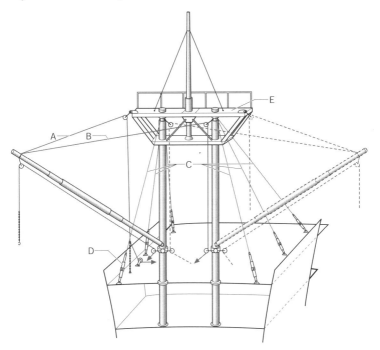

A. Fixed topping lift. B. Mobile topping lift. C. Guys. D. Stretching screws. E. Top.

— substantial and at least 1 m high on each side that is not in use for vehicle access;

— hinged or retractable on the sides used for access;

— interlocked so that the platform cannot be moved unless all the barriers are closed;

— arranged so that they cannot be opened unless the platform is at that level;

— as close to and above the edge of the opening as is practicable, so that they cannot be closed if any part of a vehicle or its cargo overlaps the deck opening;

— painted in alternate yellow and black warning stripes.

15. A flashing warning light, preferably yellow, should be fitted on the deck side of each cargo lift opening, at a place where it can be readily seen from any vehicle on the deck. The light should operate continuously when the platform is away from the opening in that deck.

16. Some ships carry mobile lifting appliances such as lift trucks and mobile cranes that can be used for cargo handling. These should comply fully with the requirements for similar equipment ashore.

4.3.2. Shore cranes

1. Care should be taken to ensure that cranes are designed for the type of application for which they are deployed and have an appropriate fatigue life. The modern method of achieving this is by giving the crane a classification based on the designer's criteria for the use of that crane. Guidance on crane classification can be found in ISO 4301 *Cranes and lifting appliances*. National and international standards may specify the requirements for new cranes in ports.

2. Automatic audible and visual alarms that operate whenever the travel motion of the crane is engaged should be fitted to the crane. The audible alarm should be distinct from any other alarm, and loud enough to warn any persons who may be in the vicinity of the wheels of the crane. The visual alarm should be a flashing light, normally of amber colour.

3. The crane should be fitted with a separate horn or similar warning device and a flashing light that can be operated manually to warn or attract the attention of any person nearby.

4. The track of a rail-mounted crane should be:

— of adequate section and bearing capacity;

— firm and level, with an even running surface;

— electrically bonded and earthed.

5. Shock-absorbing buffers should be provided on rail-mounted cranes and end stops on rails.

6. Rail-mounted cranes should be so designed that in the event of breakage of a wheel, failure of an axle or derailment, the crane will not overturn or collapse.

7. Rail-mounted cranes should be equipped with devices to clear the rails of dunnage and similar material automatically as the crane moves.

8. The wheels of rail-mounted cranes should be provided with guards to prevent danger to feet (figure 35).

9. Anemometers should be fitted in the most exposed position of large rail-mounted cranes to provide warning of wind conditions requiring them to be taken out of service. The warning should be given to the crane operator and repeated at ground level for the benefit of supervisory personnel.

10. Rail-mounted cranes taken out of service in high winds should be secured when necessary. Securing devices should be designed for the purpose (figure 36). The usual type is a storm pin or bolt that can be inserted into a socket in the quay surface. Other types include rail clamps, wheel scotches and chains.

Figure 35.   Rail-mounted crane wheel protection

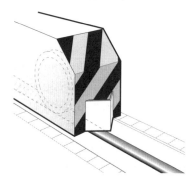

11. Arrester brakes should be fitted to large rail-mounted cranes that are liable to be exposed to high winds during use if the travel motor brakes cannot prevent inadvertent movement along the track in such circumstances.

12. Rail-mounted cranes where the distance between the gantry legs is more than 30 m should be equipped with means of synchronizing the motors to prevent any leg moving out of unison with the others.

13. If a number of rail-mounted gantry cranes working on the same track can be brought close together or come into contact with a ship's superstructure, suitable sensors should be provided to prevent them striking each other.

14. Any trapping points between a crane's flexible power cable and winding drum should be guarded, unless the drum is so placed as to be as safe as if it were guarded.

15. Old models of scotch derrick cranes may have only one motor driving both the hoisting and the derricking drums.

Figure 36.   Storm pin on rail-mounted crane

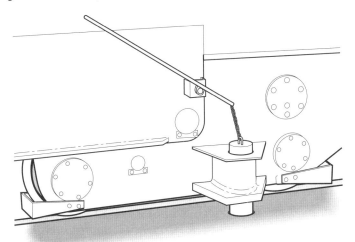

In order to avoid the possibility of interlock failure that may lead to an accident, it is recommended that such drive systems be replaced so that each motion has its own motor and brake.

16. Cranes used for lifting containers should be fitted with devices that indicate when the container spreader is correctly lowered onto the container and when the twist-locks are fully engaged and released.

17. Cranes used for lifting containers should be fitted with interlocks that prevent:

— twistlock movement, unless all four twistlocks have entered the corner fittings;

— lifting, unless all four twistlocks are fully locked or unlocked;

— twistlocks unlocking when a load is suspended from them;

— twistlock movement on a telescopic spreader, unless the frame is at the correct length;

— telescopic spreaders telescoping, unless all four twistlocks are unlocked and clear of the load.

18. Cranes used for lifting containers should be fitted with load-indicating devices that show the mass of the load being lifted.

19. The limits of stability of straddle carriers should be determined in accordance with ISO 14829 *Freight containers – Straddle carriers for freight container handling – Calculation of stability*.

20. General-purpose mobile cranes are used in many industries. However, it should be appreciated that they are designed to a relatively low classification and should not normally be used intensively for long periods of time without consultation with the manufacturer or other design authority, which may well recommend a reduction of rated capacity for such applications.

21. The chassis of crawler cranes should be clearly marked so that the operator can see the direction of travel at a glance.

22. Mobile harbour cranes should only be used on well-prepared flat ground capable of supporting them and their load. Any slope on which they travel should be within the limits specified by the crane manufacturer.

23. Great care should be taken when adding or removing lattice strut jib sections. This should always be carried out in accordance with the manufacturer's instructions, with the jib adequately supported. Persons should never be underneath the jib.

24. Every lifting appliance fitted with outriggers should be fitted with a device to indicate to the operator whether the appliance is level.

4.3.3. Lift trucks

1. When lift trucks (figure 37) are selected, it should be clearly understood that trucks powered by internal combustion engines carry flammable fuel, produce exhaust gases with toxic components and can create noise nuisance. Trucks to be used in ships' holds or other confined spaces should preferably be electrically driven.

2. Every truck driven by an internal combustion engine should:

— have an efficient exhaust system fitted with a silencer and a gas cleaner, where appropriate;

— carry an appropriate fire extinguisher.

3. The forks of lift trucks should be designed to prevent their accidental unhooking or lateral displacement when in use.

4. The forks of a truck are items of loose gear and should be tested and certified before being brought into use.

5. Trucks should be fitted with devices to automatically limit the upward movement of the forks, and, unless it is non-powered, the downward movement.

Figure 37. One type of lift truck (driver protection on mast omitted for clarity)

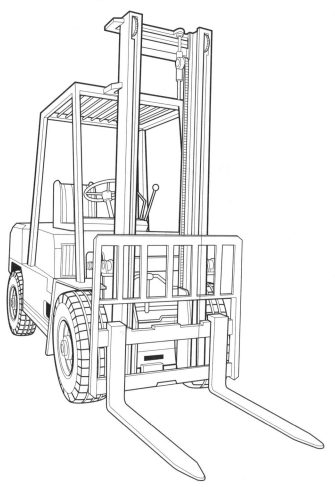

6. Any trapping, crushing or shearing points within reach of the operator in the normal operating position on the truck should be suitably guarded (figure 38).

7. All trucks and battery containers on electric trucks that are intended to be hoisted aboard ship should have suitable slinging points.

8. The steering system of trucks fitted with non-powered steering should be designed, so far as is practicable, to prevent the operator's hands from being injured if one of the truck wheels strikes a kerb, dunnage or other fixed object.

9. Every prototype or modified truck should have been stability-tested by a competent person in accordance with a national or international standard before being taken into use.

10. All trucks should be fitted with a manual audible warning device, an automatic audible warning device that operates during reversing movements, two headlights, rear lights, parking lights and reflectors, in accordance with national standards, even if they are not expected to leave the port area (figure 39).

11. Where possible, the reversing movement warning device should give a distinguishable sound that is standard throughout the port.

12. Headlights should throw forward a yellow or white non-dazzling light. Rear lights (two, as near as possible to the extremities of the vehicle in the case of wide vehicles) should throw a red beam backwards. All lights should be visible from 150 m away on a clear night.

Figure 38.   Protection for lift truck operator

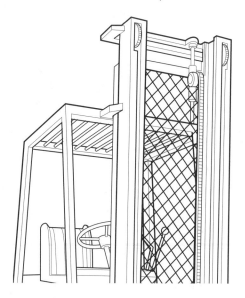

13. Two reflectors should be fitted as near as possible to the extremities of the vehicle. These should reflect red light visible from 100 m away when illuminated by headlights.

14. As lift trucks generally move both in reverse and forward, the provision of appropriate headlights and work lights is recommended. Larger vehicles should be fitted with additional reflectors at the front and sides.

15. All trucks should be fitted with flashing orange or amber lights.

16. Trucks with an enclosed cab should be fitted with one or two rear-view mirrors.

Figure 39.   Warning devices to be fitted to moving lift trucks

17. Centre-seated counterbalanced trucks should have an operator restraint system fitted to prevent the operator from being thrown from the vehicle or trapped by the

overhead guard if the truck tips over laterally. This may be an enclosed cabin, seat belt or other device.

18. An upholstered suspension seat should be fitted to minimize the transmission of shock loads to the operator and avoid compression of vertebrae. Good seats should have seat backs giving good support to the operator, but not obscuring the view to the rear of the truck.

19. Forklift trucks should be fitted with a substantial overhead guard sufficiently strong to protect the operator as far as possible against the impact of objects falling from above (see figure 38). In some cases, an additional guard to protect against small falling objects may be necessary. This may be a solid or perforated metal sheet.

20. Side windows should be automatically locked in the closed position on some types of truck, e.g. sideloaders, in order to prevent injury to the head of an operator leaning out of the cab while lowering a load.

21. All moving parts within the operator's reach should be securely guarded.

22. A load backrest extension is recommended if the truck handles small unsecured loads, e.g. small heavy boxes.

23. Sheet metal side and front guards protecting operators of platform trucks should be of a shape that enables them to climb on and off easily and quickly.

24. Lift trucks should be fitted with service and parking brakes that comply with national or international standards.

25. All trucks should be marked with their safe working load or loads (where there is more than one load owing to the use of devices such as stabilizers or extension forks)

and related load centre (figure 40). The load plate should show the safe working load of the truck at various load centres and lift heights.

26. All trucks should be marked with the unladen weight of the truck.

27. Electric trucks should be marked with their weights both with and without the battery and battery container. The battery container should be marked with the total weight of the container and battery.

28. All trucks should be provided with:

— a builder's plate giving the authorized gross laden weight, machine type and the maker's name and address;

— an operating plate giving the owner's name and address and, if possible, maintenance particulars such as servicing dates.

29. No further weight should be added to a counterweight for the purpose of increasing the lifting capacity.

30. The operating platforms of end-controlled powered trucks and tractors should be provided with substantial guards to prevent the operators from being crushed in the event of collision with obstacles or other vehicles.

31. Electrically driven trucks should be fitted with at least one adequate mechanical brake and a mechanically operated current cut-off that comes into operation automatically when the operator leaves the vehicle. When the vehicle is stationary, it should not be possible to close the circuit unless the controller has passed through the neutral position.

149

Safety and health in ports

Figure 40.   Load plate of a lift truck

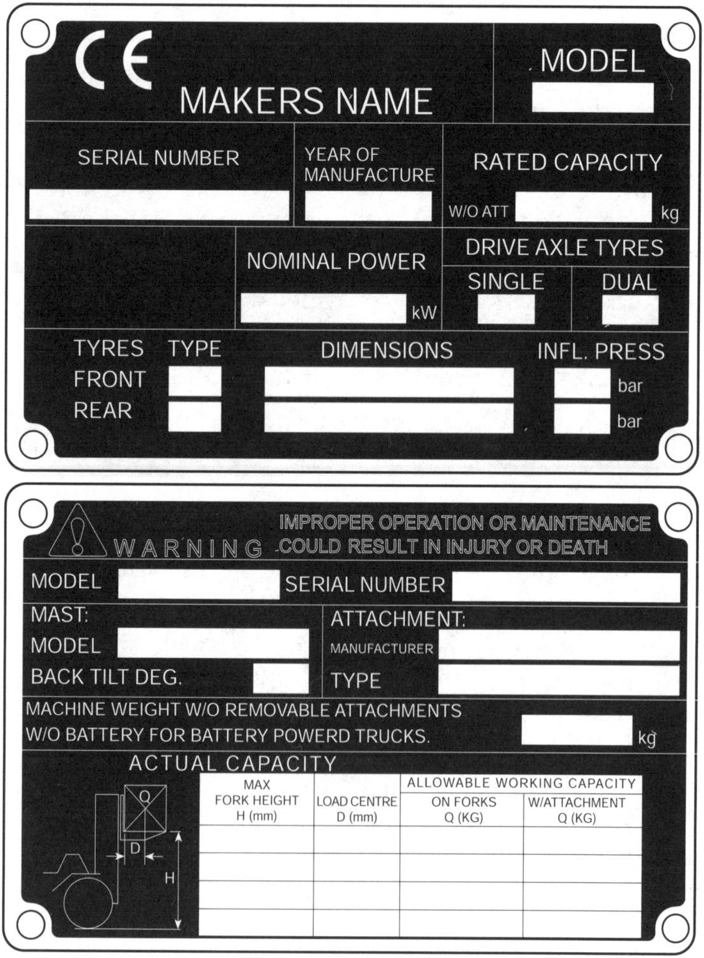

CE

MAKERS NAME

MODEL

| SERIAL NUMBER | YEAR OF MANUFACTURE | RATED CAPACITY |
| | | W/O ATT   kg |

| | NOMINAL POWER | DRIVE AXLE TYRES |
| | kW | SINGLE | DUAL |

TYRES	TYPE	DIMENSIONS	INFL. PRESS
FRONT			bar
REAR			bar

⚠ WARNING   IMPROPER OPERATION OR MAINTENANCE COULD RESULT IN INJURY OR DEATH

| MODEL | | SERIAL NUMBER | |

MAST:
MODEL
BACK TILT DEG.

ATTACHMENT:
MANUFACTURER
TYPE

MACHINE WEIGHT W/O REMOVABLE ATTACHMENTS
W/O BATTERY FOR BATTERY POWERD TRUCKS.   kg

ACTUAL CAPACITY

	MAX FORK HEIGHT H (mm)	LOAD CENTRE D (mm)	ALLOWABLE WORKING CAPACITY	
			ON FORKS Q (KG)	W/ATTACHMENT Q (KG)

32. Measures should be taken to prevent spillage of battery electrolyte, including using the minimum necessary quantity of distilled water, supporting the batteries on shock absorbers and providing drip trays.

33. Wherever possible, wheels should be positioned within the truck body. If they are positioned outside, the wheel guards should be conspicuously marked.

34. All lift trucks should be painted in a bright colour that is highly visible against the backgrounds where they operate. The back ends of rear-wheeled steered trucks should be painted in yellow and black stripes to warn of the dangers of the swinging back when manoeuvring.

4.4. Loose gear

4.4.1. General requirements

1. Wrought iron should never be used in the manufacture or repair of any loose gear.

2. Any gear made wholly or in part of wrought iron should be scrapped as soon as is practicable.

3. While any gear made wholly or in part of wrought iron remains in use, it should be periodically heat-treated in accordance with Appendix G.

4. A block should not be subjected to any form of heat treatment.

5. Every steel part of loose gear (other than wire rope) should be made of the same quality grade of steel.

6. Any welding in the manufacture or repair of loose gear should be carried out by qualified workers using appropriate techniques.

4.4.2. Chains and chain slings

1. Chains and chain slings should generally be constructed from steel bars of at least 10 mm diameter for Grade M chain and 7 mm for Grade T chain.

2. Chains that are to be used at temperatures below about -5°C should be made of special steels. Grade T chains can be used with no reduction of their safe working load at temperatures between –30° and +200°C.

4.4.3. Wire ropes and slings

1. Wire ropes should be of adequate strength for the frequency and type of intended use (figure 41), and selected in accordance with ISO 4308 *Cranes and lifting appliances – Selection of wire ropes*.

2. The guaranteed minimum breaking load should not be less than the product of the safe working load and a factor of safety determined in accordance with Appendix E.

3. Hoisting ropes should be in one length without any joins. If the lengthening of a cable is unavoidable, it should be done by an approved method, such as fitting a thimble and shackle or a Bordeaux connection. In such cases, the safe working load should be reduced by an appropriate amount. It may also be necessary to fit larger sheaves if the connection needs to pass over them.

4. Wire rope slings may be endless, i.e. formed by jointing the two ends of the rope, or have a variety of terminations and splices (figure 42).

5. A wire rope should be properly terminated.

6. Capping and splicing are skilled operations that should only be carried out by workers having the necessary expertise.

Figure 41. Construction of wire ropes

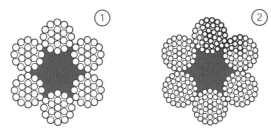

1. Rope with 6 strands of 19 wires (1+6+12) and textile core.
2. Rope with 6 strands of 37 wires (1+6+12+18) and textile core.

7. If a particular method of splicing is prescribed by national legal requirements, only that method should be used.

8. All thimble or loop splices should have at least three tucks with a whole strand of rope, followed by two tucks with half the wires cut out of each strand (figure 43). All tucks other than the first should be against the lay of the rope. If another form of splice is used, it should be equally efficient.

9. No splice, however well made, can equal the strength of the original rope. The strength of the splice gradually decreases with diameter. At the largest sizes, it may be only 70 to 75 per cent as strong as the original rope. This loss of strength should be taken into account when the factor of safety is decided.

10. A splice in which all the tucks are with the lay of the rope (Liverpool splice) should not be used in the construction of a sling or in any part of a lifting appliance where the rope is liable to twist about its axis, even if the splice is protected by a swivel.

153

Figure 42.   Steel wire slings

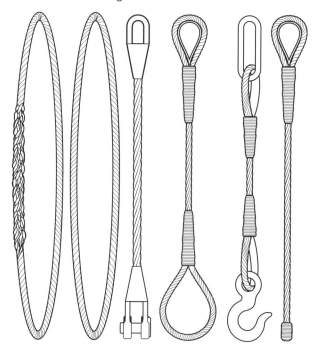

Figure 43.   Loop spliced on thimble

11. Any protection on a splice in a wire rope to a lifting appliance should only be provided at its tail. This allows any deterioration of the splice (i.e. broken wires) to be seen.

12. Compressed metal ferrules should be made to a manufacturer's standard:

— the material used should be suitable in particular to withstand deformation without any sign of cracking;

— the correct diameter and length of ferrule should be used for the diameter of the rope;

— the end of the rope looped back should pass completely through the ferrule;

— correct dies should be used for the size of the ferrule;

— correct closing or compression pressure should be applied to the dies;

— tapered ferrules, where the end of the rope is not visible for inspection after closing, should not be used.

13. Terminal fittings on wire ropes should be capable of withstanding the following minimum loads:

Diameter of rope	Percentage of rope's minimum breaking load
Up to 50 mm	95
More than 50 mm	90

14. A wedge socket used as a terminal fitting of a lifting appliance should be suitable for the size of rope and be properly fitted.

15. The tail of the rope should protrude sufficiently from the socket to enable it to be bent back upon itself to form a loop, and for the end then to be clamped or lashed to

itself after emerging from the socket (not clamped to the main part of the rope).

16. The wedge should be inserted and driven home by gentle hammering with a mallet.

17. A heavy load (up to the safe working load of the socket, if this is practicable) should be lifted a short distance and then be allowed to descend and be braked normally in order to bed the wedge.

18. A Lang's lay rope (also known as a Liverpool splice) should only be used if it is not free to twist about its axis, i.e. both ends of the rope are secured (figure 44).

19. Bolted clamps (such as Crosby, plate or bulldog grips) should not be used to form a terminal join in any hoist rope, derricking rope, guy of a ship's derrick or derrick crane, or in the construction of a sling (figure 45).

20. A rope made of fibre interspersed with wire strand should not be used on a lifting appliance such as a crane, but may be used as a sling in certain circumstances, subject to testing in accordance with Appendix B and certification on the basis of a factor of safety in accordance with Appendix E.

21. Before a wire rope is put into service, it is essential to verify from tables or calculations that it is of the correct diameter for winding on winch ends or sheaves. The winding diameter should generally be at least four times the cir-

Figure 44.   Lang's lay rope/Liverpool splice

Figure 45. A bolted clamp

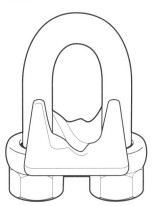

cumference of the rope (practically 12 times its diameter), but it is advisable to use higher ratios. The following rules are frequently adopted:

— for slow-moving appliances, the diameter of pulleys and sheaves should be 300 times the diameter of the thickest wire in the rope, and for most lifting appliances 500 times that diameter;

— the diameter of pulleys and sheaves should also be at least 24 times the diameter of a rope with 6 H 37 wires, and at least 20 times the diameter of a rope with 6 H 61 wires.

4.4.4. Fibre ropes and slings

1. Natural fibre rope for use on a lifting appliance or for slings should be of good grade manila (abaca), sisal (aloe) or other fibre of equal quality manufactured to a national or

international standard, or in accordance with the requirements of a classification society.

2. Natural fibre slings are usually manufactured from three-strand rope. The splice should be dogged off or a tail allowed. Natural fibre slings are usually made with soft eyes or are endless.

3. As natural fibre ropes are affected by damp, it may be advantageous to use ropes that have been treated with a suitable rot-proofing and/or a water-repellent agent.

4. A thimble or loop spliced in a natural fibre rope should have not less than four full tucks, with all the yarns in the strand tucked against the lay. The splice should then be dogged.

5. A synthetic fibre rope should not be used as a sling or as part of a lifting appliance unless:

— it is made to a recognized national or international standard, or in accordance with the requirements of a classification society;

— the manufacturer has certified its guaranteed minimum breaking load;

— its diameter is more than 12 mm.

6. A synthetic fibre rope should not be:

— used on a pulley block that does not meet the requirements of section 4.4.5;

— reeved through a pulley block on which:

  • the groove of a sheave is less than the diameter of the rope; or

  • the sheave has any defect likely to cause damage to the rope.

7. A synthetic fibre rope intended to be used for lifting should not be spliced to a natural fibre rope.

8. When a synthetic fibre rope is joined to a wire rope, the two ropes should have the same direction of lay. A thimble should be fitted to the eye of the fibre rope and the ropes shackled together.

9. Synthetic fibre rope slings are usually manufactured from three-strand rope and spliced in the same way as natural fibre slings. The fibre can be indicated by the colour of the identification label as follows:

— green – polyamide (nylon);
— blue – polyester (terylene);
— brown – polypropylene.

10. A thimble or loop splice:

— in a polyamide and polyester fibre rope should have at least four tucks with all the yarns in the strands, followed by one tuck with approximately half the yarns of each strand, and a final tuck with at least one-quarter of the yarns;

— in a polypropylene fibre rope should have at least four full tucks, with all the yarns in the strands.

11. All tucks should be against the lay of the rope.

12. Tails protruding from the rope should be at least three rope diameters long or be dogged.

13. Synthetic fibre webbing slings for general use should be at least 35 mm and not more than 300 mm wide. Specially designed slings may be wider. Slings can be manufactured endless or with soft eyes. The eyes of slings over 50 mm wide are reduced by folding at the time of manufacture to allow

them to be accommodated in hooks and shackles of the correct safe working load. The eyes may be fitted with reinforcing at point-of-hook contact. Wear sleeves may also be fitted to reduce damage to the main body of the slings.

14. The minimum length of a soft eye measured internally when the webbing is laid flat should be:

— three times the width of webbing up to 150 mm wide;

— two-and-a-half times the width of webbing, for widths greater than 150 mm.

15. Any substance used to increase the resistance of a webbing sling to abrasion should be compatible with the synthetic fibre.

16. Polypropylene webbing or rope slings likely to be exposed to prolonged bright sunshine should be manufactured of material stabilized against degradation by ultra-violet light, as otherwise severe loss of strength may occur in a relatively short period.

17. The stitching material should be of the same synthetic yarn as the sling, and the join should be such that, so far as is practicable, the load is distributed equally across the width of the belt.

18. Webbing slings should be manufactured to an internationally or nationally recognized standard, supported by an internationally recognized quality management system.

19. Disposable or one-trip slings should:

— be not less than 25 mm wide;

— have a breaking load at least five times their safe working load if up to 50 mm wide and at least four times their safe working load for wider slings.

20. Disposable or one-trip webbing slings should be clearly and durably marked in a suitable place with the following:

— the safe working load at angles from 0° to 45° from the vertical;

— either the mark "U", indicating a disposable sling, or the word "disposable" or "one way" in English;

— the maker's identification mark;

— the batch number relating to the test certificate or certificate of conformity of the sling;

— the year of manufacture.

21. Round slings should not be used for cargo handling.

4.4.5. Blocks

1. Pulley blocks for use with synthetic or natural fibre ropes should have a cast housing or side and partition plates, and straps of steel or of wood suitably reinforced with steel, or aluminium straps.

2. Except in the case of a cast housing, the side straps should be adequately and properly secured to the head fitting.

3. The diameter of the sheave(s) measured at the bottom of the groove should not be less than 5.5 times the design rope diameter.

4. The rope groove should have a depth of not less than one-third the diameter of the rope and a radius of not less than 1 mm greater than half the diameter of the rope.

5. A block should generally not be fitted with more than three sheaves and a becket (figure 46), or four sheaves if the block has no becket.

Figure 46.   Three-sheave block with becket

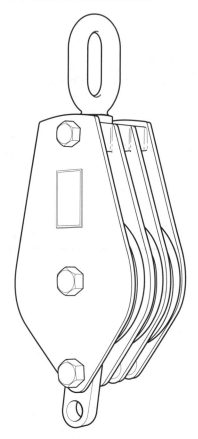

6. Provision should be made for the lubrication of all metal bearings and swivel-head fittings and, where necessary, any plastic bearings.

7. The safe working load of the block should be based on use with best grade manila rope.

8. The block should be marked with:

— the size of manila rope for which it has been designed;

— its own safe working load;

— its identification marks.

9. The safe working load of a single-sheave block is the maximum load that can be safely lifted by that block when it is suspended by its head fitting and the load is secured to a wire rope passing round its sheave.

10. When a single-sheave block is rigged with the load to be lifted secured to its head fitting and the block is suspended by a wire rope passing around its sheave, it should be permissible to lift a load twice the safe working load marked on the block (figure 47).

11. The safe working load of a single-sheave block incorporated elsewhere in a derrick rig that is secured by its head fitting and subjected to tension arising from a wire rope that forms part of the derrick rig and passes around or partially around the sheave is half the resultant load upon its head fitting. Allowance should be made for the effects of friction in the block and rope stiffness, i.e. the extra load arising from the effort of bending the wire rope partially around the sheave.

12. The safe working load of a multi-sheave block is the maximum force that may be applied to its head fitting.

13. The design of blocks to be used with wire ropes should be based on a wire rope having a tensile strength of 180 to 200 kg/mm^2 (1,770 to 1,960 N/mm^2).

Figure 47.   Safe working load of a single-sheave block

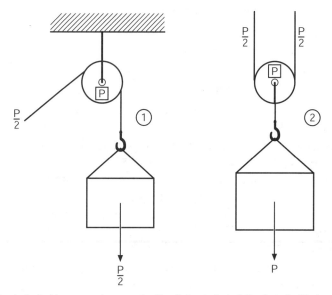

1. Load attached to rope passing around pulley. 2. Load attached directly to the block.
P. Safe working load of the block.

14. A cargo block fitted to the heel of a derrick for the cargo runner or hoist rope should be provided with a duck-bill or similar type of head designed to restrict the downward movement of the block when the runner becomes slack.

15. A cargo block fitted to the head of a derrick used in union purchase, and in other cases where practicable, should be fitted with a swivelling eye.

16. Cargo blocks should be rigged in accordance with the ship's rigging plan.

4.4.6. Other loose gear

1. Hooks should be constructed so as to cause as little distortion and damage to the eye of a sling as possible. The larger the hook that can be used, the less distortion is caused to the sling.

2. Every hook should be provided with an efficient device to prevent the displacement of the load from the hook, or be of such construction or shape as to prevent displacement (figure 48). These may be safety latches, "C" hooks, ring assemblies for union purchase or ramshorn hooks for use with heavy lifts.

3. The screwed shank of a hook or other similar thread should be undercut to a depth no greater than that of the

Figure 48. Safety hooks with two types of latches

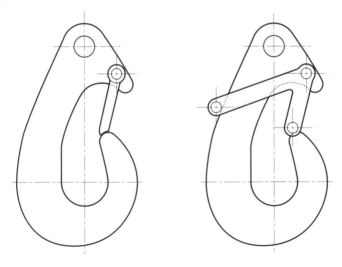

thread. Any corner where a plain portion of the shank terminates at a shoulder or flange of greater diameter should be radiused as far as is practicable.

4. Hooks may be attached to chain slings by mechanical connection or shackled to slings of any material, or may be an integral part of a block.

5. The shape of thimbles should be such that the internal length and width are six and four times the diameter of the rope respectively (figure 49). The thickness of the metal at the bottom of the throat should be 0.4 times the rope diameter.

6. The diameter of the body and pin of a shackle are given (figure 50), as well as its safe working load (13 mm (1/2") H 16 mm (5/8")). Shackle pins are always larger than the body of the shackle.

7. Shackles are usually manufactured from two types of steel, grade T (800 N/mm^2) and grade M (400 N/mm^2). T shackles are approximately twice the strength of grade M shackles. They are usually known as alloy and high tensile (HT) shackles. Sizes of different types of shackles are as follows:

Size	High tensile (t)	Alloy (t)
13 mm (1/2")	1.0	2.0
25 mm (1")	4.5	8.5
50 mm (2")	19.0	35.0

8. Where shackles are permanently rigged, the pins should be locked by mousing a screw collar pin or by a split cotter pin on a nut and bolt pin.

Figure 49.   Thimble

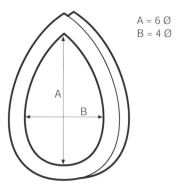

A = 6 Ø
B = 4 Ø

Figure 50.   Shackle

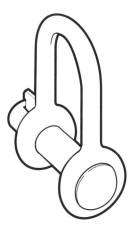

167

9. A swivel should always be inserted between the sling hook and the hoisting rope.

10. Every vacuum lifting device should be fitted with the following:

— a suitable vacuum gauge or other device clearly visible to the lifting appliance operator;

— an audible warning to the operator and any person working nearby when the vacuum is 80 per cent or less of the designed working vacuum, or the vacuum pump ceases to operate;

— means for maintaining a sufficient vacuum to continue supporting the load for sufficient time to allow it to be lowered safely from the maximum height of lift of the lifting appliance to the quayside in the event of vacuum pump failure.

11. The vacuum gauge should be marked in red with the lowest vacuum at which the appliance may be used.

12. The designed working vacuum should be the vacuum necessary to support the test load which the lifting appliance is required to support.

13. Where the vacuum is controlled from the cab of the lifting appliance, the controls should prevent accidental removal of the vacuum.

14. As far as is practicable, the surface of a test load of a vacuum lifting device should be similar to the worst type of surface the device is intended to lift. If the lift is to be wrapped, the test load should be similarly wrapped.

15. The voltage of the electric power supply to any magnetic lifting device should not fluctuate by more than ±10 per cent.

16. A magnetic lifting device should be:

— provided with an alternative power supply unless the magnet is used only to handle scrap metal or other cargo, and no person will be near the device;

— constructed to withstand the entry of moisture.

17. A magnetic lifting device should be marked with its safe working load as determined by tests using weights of the same characteristics as the load for which the device is intended to be used. When the load to be lifted is dissimilar to the test load, it should be restricted to approximately 60 per cent of the safe working load.

18. Other loose gear includes lifting beams, spreaders, lifting frames and other attachments for lift trucks, tongs, claws and cradles for handling round bars or logs. All should have adequate strength for their intended purpose with an appropriate factor of safety. The effectiveness of tongs and claws depends on the roughness of their surface or the condition of their teeth.

4.5. Lifting devices forming an integral part of a load

4.5.1. General requirements

1. Lifting devices forming an integral part of a load are not loose gear, but should be:

— of good design and construction;

— of adequate strength for their proposed use;

— maintained in good repair.

2. Such devices include eye bolts, integral lifting lugs on plant, corner fittings on containers, the lifting straps of flexible intermediate bulk containers (FIBCs), and pallets secured to a load.

169

3. If part of a load is secured to a lifting device that forms an integral part of a load by further means, it is essential that this also is of adequate strength and maintained in good repair.

4.5.2. Flexible intermediate bulk containers (FIBCs)

1. Some FIBCs (for carrying powdered homogeneous cargo) are reusable, but single-trip FIBCs should never be reused.

2. The lifting straps at the corners of FIBCs should always be lifted vertically (figure 51).

Figure 51.   Flexible intermediate bulk container (FIBC)

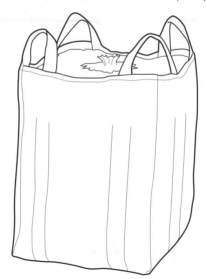

3. Before an FIBC is lifted, the certificate of conformity and a thorough examination certification (issued in the last 12 months) should be checked and the bags should be inspected.

4.5.3. Pallets

1. Pallets should be free from visible defects liable to affect their safe use (figure 52).

2. The decks of wooden shipping pallets should be at least 35 mm thick. The space between the decks should be sufficient to allow easy access by the forks of lift trucks or the arms of other pallet-lifting devices.

Figure 52.   Standard pallet

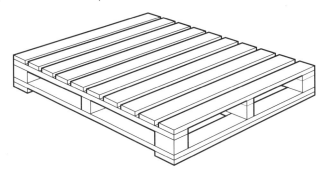

5. Safe use of lifting appliances and loose gear

5.1. Basic requirements

5.1.1. General requirements

It is essential that all who work in ports are aware of the basic potential hazards of lifting operations. To control these hazards it is necessary to ensure that:

— all lifting equipment is suitable for the proposed operation and environment;

— initial and continuing integrity of the equipment can be demonstrated;

— all personnel are appropriately trained and supervised;

— lifting operations are properly planned and managed;

— safe systems of work are followed;

— the equipment is regularly maintained.

5.1.2. Planning and control of lifting operations

1. All lifting operations should be planned and carried out under the control of a responsible person. Operators of lifting appliances should be competent to control routine operations under the general control of management, but more complex and specialist operations should be carried out under the direct control of a person with the necessary knowledge and experience.

2. Matters to be considered when planning lifting operations should include the following:

— type and size of ship and cargo;

— type of loads to be lifted;

— particular lifting hazards associated with those loads (position of centre of gravity, stability, rigidity, etc.);

— any handling symbols marked on cargo (figure 53);

— attachment of the load to the lifting appliance (availability of appropriate loose gear);

— frequency of the lifting operation;

— where the loads are to be lifted from and to;

— selection of appropriate lifting appliances;

— position of the lifting appliance (sufficient space and level ground);

— proximity hazards (power cables, buildings, roadways, other cranes, etc.);

— requirements for safe erection of the lifting appliance (space, ground loadings, level, etc.);

— ground loadings that will be applied by the lifting appliance and any necessary equipment to spread the load;

— provision of competent staff (lifting appliance operators, slingers, signallers, supervisors, etc.);

— safe systems of work for taking the lifting appliance out of service during maintenance, thorough examination, testing and repairs;

— emergency procedures, including rescue of an operator from an elevated position;

— systems for reporting breakdowns, accidents and dangerous occurrences;

— systems to prevent any unauthorized movement of lifting appliances;

— provision and maintenance of appropriate safety equipment.

Figure 53.   Cargo handling symbols

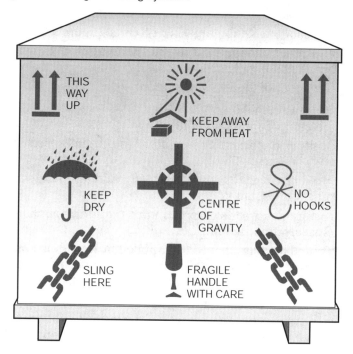

3. The planning should be constantly reviewed to ensure that any changes are adequately considered.

4. International standard ISO 12480, *Cranes – Safe use – Part I: General* gives guidance on the safe use of cranes.

5.1.3. Training

1. All lifting appliance operators and users of loose gear should be carefully selected, trained and tested to ensure that

they are competent. Operators should be trained and certified to operate each make and model of lifting appliance which they operate.

2. International standard ISO 15513 *Cranes – Competency requirements for crane drivers (operators), slingers, signallers and assessors* provides guidance on the competency requirements for crane operators, slingers, signallers and assessors. Further guidance on the training of crane operators can be found in ISO 9926 *Cranes – Training of drivers*.

5.1.4. Inspection

5.1.4.1. General inspection requirements

1. All lifting appliances and loose gear should be regularly visually inspected before and during use with a view to checking for obvious deterioration and determining whether they are safe for continued use.

2. Inspection is a completely separate process from maintenance. Inspections should be carried out by conscientious, responsible personnel. Lifting machine operators and slingers are often competent to carry out daily and weekly inspections, but checks are needed to ensure that they have the necessary competence.

5.1.4.2. Daily checks

1. All lifting appliances should be inspected at the beginning of each shift or working day during which they are to be used. The use of a checklist is recommended.

2. The checks should include, as appropriate for the type of appliance, all daily checks specified in the manufacturer's handbook, and checks to ensure that:

— all ropes are correctly positioned on their sheaves and drums are not displaced;

— electrical equipment is not exposed to contamination by oil, grease, water or dirt;

— relevant levels and/or components show no loss of fluids (e.g. lubricating oil, coolant);

— all limit switches, cut-outs and dead man's handles or levers operate correctly; caution should be taken during checking in case of malfunction;

— the safe working load limiter is correctly set and the manufacturer's daily test carried out;

— the radius indicator is appropriate to the jib configuration fitted, if separate from the safe working load limiter;

— the load-lifting attachment radius is varied without load to check the correct movement of the radius indicator and safe working load limiter;

— correct air pressure is maintained in any pneumatic control system (e.g. brakes);

— items such as lights, windscreen wipers, washers and other attachments are properly secured and operate efficiently;

— wheels are secure, and the condition and pressure of tyres are appropriate on wheel-mounted lifting appliances;

— all controls function correctly without load;

— audible warning devices operate correctly;

— the appliance is in tidy condition and free from tins of oil, rags, tools, or materials other than those for which storage provision is made;
— safe access is provided;
— appropriate fire-fighting equipment is available;
— no obstructions are present in the path of travel of a crane.

3. Appropriate records should be kept. As a minimum, these should record that the inspection has been carried out and any defects found that could not be immediately rectified. Such defects should be reported for rectification.

5.1.4.3. Weekly checks

1. All lifting appliances should be inspected once a week when in use. In addition to the items for daily inspection, the checks should include, as appropriate for the type of appliance:

— weekly checks specified in the manufacturer's handbook;
— visual inspection of all ropes for broken wires, flattening, basket distortion, excessive wear or surface corrosion, or other signs of damage;
— checks of all rope terminations, swivels, pins, retaining devices and sheaves for damage, worn bushes or seizure;
— checks of the structure for damage (including missing and bent bracings on bridges and strut jibs, as well as bulges, indentations and unusual rubbing marks, cracked welds and loose bolts or other fasteners);
— inspection of hooks and other load-lifting attachments, safety catches and swivels for damage, free movement or wear, and checks to ensure that hook shank threads

and securing nuts do not show signs of excessive wear or corrosion;

— checks to ensure correct operation and adjustment of controllers;

— inspections to identify any creep of hydraulic rams and hoses, any fitting deterioration on hydraulic machines, and any oil leaks;

— checks to ensure the effectiveness of brakes and clutches;

— inspections of tyres for damage and wear on walls and tread, and checks of wheel nuts for tightness on wheel-mounted mobile appliances;

— inspections of slew locks, if fitted;

— inspections of steering, brakes (foot and parking), lights, indicators, warning devices, windscreen wipers, and washers.

2. The results of all inspections of lifting appliances should be recorded. Details need be recorded only of defects found.

5.1.4.4. Blocks

Inspections of blocks should check that:

— sheaves are not cracked at the rim, and no part of the rim is missing;

— grooves are not excessively worn;

— sheaves turn freely and smoothly;

— head-fitting swivels are securely fastened and free from visible defects;

— shanks are not distorted, turn freely by hand and are not slack in their holes;

— clearance between sheaves and partitions of side plates is not excessive;

— side straps are sound and free from any cracks;

— greasing arrangements are satisfactory and grease nipples have not been painted over;

— data plates are intact and legible.

5.1.4.5. Equipment not in regular use

1. The extent and thoroughness of inspections of lifting appliances not in regular use before each use should be based on the length of the period that the appliance is out of use and its location during this period. An appliance standing out of use under cover or inside a workshop may only require the checks recommended in sections 5.1.4.2 and 5.1.4.3.

2. An appliance left out of use exposed to the weather, atmospheric pollution, etc., may require an extensive appraisal to determine its fitness for use. This should include:

— any checks recommended by the manufacturer;

— examination of all ropes for signs of corrosion and damage;

— examination of all control linkages for evidence of seizure or partial seizure;

— checks to ensure thorough and correct lubrication;

— testing of each crane motion for several minutes without load, first individually, then two or more motions simultaneously, as appropriate, before repeating the tests with a load;

— checking the correct functioning of all crane safety devices;

179

— checking hoses, seals and other components for evidence of deterioration.

3. Loose gear that is not in regular use should be returned to the gear store on shore or on board ship (see section 5.3.3.2).

5.1.5. Weather conditions

1. Lifting operations should only be carried out in weather conditions that are permitted in the relevant operating instructions.

2. Adverse weather conditions in which lifting operations may need to stop include:

— high winds;

— lightning;

— dangerous impairment of visibility by rain, snow, fog, etc.;

— adverse sea states;

— significant vessel movement from wash.

3. Warning of adverse weather should not rely solely on anemometers on cranes. Weather forecasts should be obtained so that appropriate steps can be taken before the arrival of the high winds or other adverse weather conditions.

4. Even at lower wind speeds, it may be dangerous to continue lifting operations, particularly when the load on a crane has a large surface area (e.g. a container). Lifting operations should stop if it is likely to become difficult to control the movement of the load.

5. Operating instructions should include the actions to be taken by specified persons in the event of adverse weather.

6. When high wind speeds are expected, cranes should be secured in their appropriate out-of-service condition. If this requires the raising or lowering of a jib, the planned procedure should ensure that there is adequate time and space to do so. Rail-mounted cranes should be secured. Cranes secured at picket points should be travelled against the wind to the nearest picket position and the storm anchor inserted.

7. Lifting operations should be stopped and all persons withdrawn from the vicinity of the crane if there is a possibility of the crane being struck by lightning.

8. A crane that has been struck by lightning should be thoroughly examined before being returned to service.

9. Ropes attached to the load (tag lines) may be used to help control loads in light winds, but it is essential to ensure that workers holding tag lines are fully aware of the motions to be performed by the crane. Workers holding such lines should never attach them to, or wrap them around, their bodies. The lines should be held so that they can be instantly released if necessary.

5.2. Lifting appliances

5.2.1. General requirements

5.2.1.1. Safe use

1. Lifting appliances should only be used in accordance with the manufacturer's instructions.

2. Operating rules incorporating safe systems of work should be drawn up for all lifting operations.

3. All movements of deck cranes controlled by limit switches should be tested before use.

4. Cranes should only lift loads vertically.

5. A lifting appliance operator should not be permitted:

— to use a limiter as the normal means of stopping a motion;

— to use a working load limiter as the normal means of determining that a load can be lifted or lowered.

6. Loads should never be dragged or moved in any manner that exerts a side load on a crane or lift truck. If it is necessary to drag a load for a short distance, for example on the 'tween decks area of a ship, a snatch block should be used (see also section 7.5.2, paragraph 14).

7. There should be a minimum 600 mm clearance between any part of a crane and any fixed object. Persons should be prevented from entering any area where the clearance is less than 600 mm.

8. All personnel not directly involved in the lifting operation should be kept clear of the area.

9. No person should stand under a suspended load.

10. No person should be lifted by a lifting appliance other than in a specifically designed personnel carrier.

11. No persons should be permitted to board or leave a lifting appliance without first obtaining the operator's permission. If the access point is out of sight of the operator, means should be provided to ensure that the operator is

aware of the whereabouts of the other person. A notice specifying the boarding procedure should be posted at the boarding point, where appropriate.

12. Lifting appliance operators should:

— only perform lifting operations when specifically instructed to do so by the designated signaller; however, every emergency stop signal should be obeyed;

— perform the operations smoothly, avoiding sudden jerks;

— ensure that the power supply is turned off before leaving the appliance.

13. Lifting appliance operators should never:

— lift loads over persons;

— leave loads suspended longer than is necessary to move them;

— leave appliances unattended with a load suspended;

— allow workers to travel with loads other than in personnel carriers.

5.2.1.2. Care and maintenance

1. All wire ropes on lifting appliances should be regularly treated with a wire rope dressing free from acid or alkali. Whenever possible, this should be of a type recommended by the manufacturer.

2. Where it is practical and safe to do so, the dressing should be applied where the rope passes over a drum or pulley, as the bending of the rope facilitates the penetration of the dressing.

3. It may be necessary to clean wire ropes used in dusty or abrasive environments thoroughly before applying the dressing.

4. Clear evidence of deterioration will often be present in the form of barbs or fins formed by broken wires. Such barbs are dangerous when ropes are handled. However, deterioration may also be due to rotting of the textile (fibre) core. This deprives the steel wire strands of all their support, and the rope then undergoes deformation, which becomes progressively more apparent.

5. If a wire rope has deteriorated, the defective parts should not be joined together.

6. Wire ropes should be replaced when:

— they show significant signs of corrosion, particularly internal corrosion;

— there is any tendency towards "bird caging" (separation of the strands or wires);

— they show signs of excessive wear indicated by flats on individual wires;

— the number of broken wires or needles in any length of ten diameters exceeds 5 per cent of the total number of wires in the rope;

— broken wires:
  • appear in one strand only;
  • are concentrated in a shorter length of rope than ten diameters; or
  • appear in the tucks of a splice;

— there is more than one broken wire immediately adjacent to a compressed metal ferrule or any compressed termination fitted in accordance with section 4.4.3, paragraph 8, concerning thimble or loop splices.

7. Further guidance on the examination of wire ropes and discard criteria is given in the international standard ISO 4309 *Cranes – Wire ropes – Code of practice for examination and discard*.

8. The reason for any defects found should be investigated and remedial action taken.

5.2.2. Ships' lifting appliances

5.2.2.1. Ships' derricks

1. When a derrick (figure 54) is rigged:

— a person should be stationed at each span winch and/or cargo winch in use;

— only persons engaged in the rigging work should be allowed in the vicinity. Other persons should only pass along the deck with the permission of the person in charge of the operation;

— wire ropes should be checked to ensure that they are free from corrosion, kinks, needling or other patent defects;

— all shackles and securing blocks should be fitted correctly, with their pins properly tightened and secured by seizing with wire or other effective means;

— block sheaves should be checked to ensure that they are free to turn and properly lubricated;

— guys, including preventer guys where appropriate, should be properly attached to the derrick head and the correct deck eye plates in order to prevent possible jack-knifing;

— it is essential to ensure that the gooseneck is free to swivel. This may be done when the derrick is at a low angle, from 30° to 50°, with one or more persons gently swinging on the guy(s);

— a heavy lift derrick should be checked to ensure that any temporary mast or Samson post stays are properly fitted and that any special slewing guys directly attached to the lower cargo block are properly rigged;

— rigging items should not be able to whip against the winchman.

2. When deck cargo stowed on a ship makes the deck eye plates inaccessible, the guys should be secured to wire rope or chain pendants designed specially for the purpose. The pendants should be sufficiently long to enable the guys to be coupled to the pendants at the top level of the deck cargo. Extreme care should be taken to ensure that the relative positions of the guys remain as shown on the rigging plans.

3. No derrick should be rigged and adjusted for angle other than by its own power-operated topping winch or by a span winch.

4. When a topping winch is used, a person should stand by the pawl-operating gear ready to engage the pawls when signalled by the person hauling in or paying out the rope whip.

5. It is advisable to use separate drums for the luffing rope and the topping lift.

Figure 54. Derricks in union purchase rig

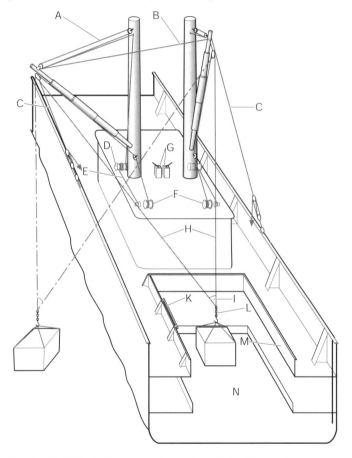

A. Topping lift. B. Tie. C. Pendant. D. Topping drum. E. Auxiliary topping rope.
F. Hoisting winches. G. Winch couplers. H. Hoisting ropes. I. Triangular piece for joining
the two hoisting ropes. K. Roller for protection of hoisting ropes (optional). L. Chain with
swivelling hoisting hook. M. Hatch coaming. N. Hatch.

187

6. No attempt should be made to engage the pawls while the winch drum is rotating in the direction for lowering the derrick.

7. A whip rope used for driving a topping winch should:

— not be used on a drum that is liable to damage the rope;

— not have more turns on the drum than is necessary for safety, provided that extra turns should be made on a drum that is whelped;

— not be surged or rendered on the drum, particularly in the case of synthetic rope; frictional heat is liable to damage the rope;

— not contain any splice;

— be of a suitable size to ensure adequate strength and handling.

8. The power of winches should be limited to a value corresponding to the safe working load of the derricks. This is likely to range from 18.6 to 37 kW (25 to 50 hp) for hoisting speeds of 0.4 m/s for 8 t loads, 0.6 m/s for 3 t loads.

9. The winch operators should:

— be protected against the weather, preferably by a sheet metal cab with large windows;

— have a clear view of the hatch, unobstructed by steam or otherwise;

— wear suitable gloves to protect hands from possible burns;

— coil rope on the deck when appropriate;

— never stand in any bight of the rope.

10. A chain stopper should not be used on derrick spans.

5.2.2.2. Use of coupled derricks (union purchase)

1. The arrangement known as "union purchase", or "married falls", allows the load to be moved sideways over the deck without slewing the boom. This may be done by using two booms or by using one boom and a fixed point, possibly on a building, at right angles to the hold.

2. When two booms on the same mast are used, union purchase is generally used for light loads of not more than 3 tonnes.

3. Calculation of the stresses on the various parts of the system should be carried out by a competent person.

4. The angle between the two cargo falls should not ex-ceed 90° at any time. As the angle increases above 90°, the stresses on the ropes and booms increase rapidly. The tension in each of the ropes is –

$$\frac{P}{2 \cos \alpha}$$

where P is the weight of the load lifted in tonnes and α is the angle between the cargo runner and the vertical (figure 55).

The table shows the values of the stress –

$$\frac{1}{2 \cos \alpha}$$

– with variations in the angle of lift α P weight of load:

α	$\dfrac{1}{2 \cos \alpha}$
10°	0.508
20°	0.532
30°	0.577
40°	0.653
50°	0.778
60°	1.000
70°	1.461
80°	2.800

5. The load in union purchase should generally be limited to half the safe working load of the weaker of the two booms used.

6. Before any union purchase operation is undertaken, reference should be made to the union purchase certificate

Figure 55.   Tension in union purchase cargo falls

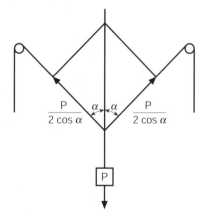

and rigging plan for the operation. Particular attention should be paid to the position of the deck lugs for the guys and preventer guys; these should be specially marked.

7. The guys holding the booms should be so placed that their horizontal projection is approximately in the plane of the load's travel.

8. The derricks should not be coupled until preventers have been put in place, unless the guys and other permanent rigging have been specifically planned for union purchase.

9. When a derrick is to be used in union purchase:

— a preventer guy should be fitted in addition to the main guy, care being taken not to confuse a guy intended only for trimming a boom with a working guy;

— the preventer and the main guy should be attached to deck eye plates that are separate but placed as close together as practicable;

— the preventer and the main guy should be adjusted when the derrick boom is under a slight dynamic loading, such as a suspended heavy hatch beam;

— the main (working) guy should be under slightly more tension than the preventer guy.

10. Where the length of a guy is adjusted by a claw device in conjunction with a series of metal ferrules compressed to a wire rope secured to a deck eye plate, the claw should be of suitable design and of adequate strength, and arranged so that it will not be accidentally released in the event of temporary partial slackness in the guy. If a fibre rope block and tackle is used, the rope should be of synthetic fibre, as this has better elasticity and does not need adjustment when it becomes wet or dry.

11. The hoist ropes of the two derricks should be secured by the use of an equalizing pulley or by suitable swivels to a common ring carrying the cargo hook (figure 56). The hook should be fitted as close to the junction of the falls as possible.

Figure 56.   Equalizing pulley for two coupled cranes

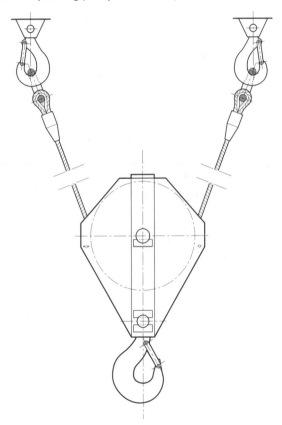

12. When derricks are in use in union purchase:

— the load should be raised just enough to clear the coaming, bulwark or railings, whichever is the highest;

— slings on loads should be of minimum length to enable the height of lift to be kept as low as possible.

5.2.2.3. Ships' cargo lifts

1. A scissor lift should be provided with temporary fencing on any side of the lift from which it is not being loaded or unloaded at that time (figure 57).

Figure 57.   Ship's cargo lift (other safety features omitted for clarity)

2. No person, other than the operator, the operator of a vehicle or persons loading or unloading the platform, should be allowed near the lift when it is in use.

3. No person, other than the operator of a vehicle who remains at the controls of the vehicle, should travel on a cargo lift platform.

5.2.2.4. Ships' mobile lifting appliances

The layout of the controls of mobile lifting appliances, such as lift trucks and mobile cranes that belong to a ship (see section 4.3.1, paragraph 16), should be checked before the appliances are used. If the layout is different from those on similar equipment on shore, operators should receive familiarization training before using them and take particular care to prevent unexpected motion.

5.2.2.5. Cranes temporarily installed on ships

1. The effects of possible list and movement of a ship, barge or pontoon should be considered when a shore crane is placed on board. These may adversely affect the strength and stability of the crane or the operation of its motions, and make it necessary to restrict the load which the crane may lift. Where there is any doubt, advice should be sought from an appropriate crane design authority.

2. A complete design assessment of the installation should be carried out if the crane is to remain on board for an extended period. This should take into account the means of securing the crane. After such an assessment, the crane should be tested to ensure adequate stability, adequate freeboard and the correct ratings.

5.2.3. Shore cranes

1. There should be a minimum clearance of 1 m between a rail-mounted crane and any obstacle it passes, including stacked goods or a vehicle being loaded or unloaded. If goods are permanently stacked near a crane track, the boundary of the stacking area should be conspicuously and permanently marked on the ground.

2. If it is not practicable to provide and maintain a clearance of 1 m, effective steps should be taken to prevent access by persons.

3. The track of a rail-mounted crane should be kept clear of loose material and rubbish as far as is practicable.

4. Travel routes to be used by mobile cranes should be checked for level, and to ensure that they are able to take the wheel loads and that there is sufficient overhead clearancc from pipcs, cablcs and other hazards. Inclines and cross-falls on the route should be checked if cranes are to travel with the jib extended or elevated.

5. If the ground is not capable of withstanding the weight of a rubber-tyred crane and its load, packing should be placed beneath the jack pads to spread the loads over an area sufficient to provide adequate support and prevent the crane from overturning or becoming unstable. Packings should be suitable for the purpose. A bed of sand may ensure more even distribution of the loads and prevent damage to the packing material.

6. It is essential to ensure that the chassis of a "free-on-wheels" mobile crane is level before use.

7. Outriggers should always be used in accordance with the manufacturer's instructions. Cranes should never be used with outriggers extended only on one side.

8. Cranes that are out of service overnight or longer should be left in the condition specified in the manufacturer's instructions.

9. A number of accidents have occurred when the bow or stern of a ship approaching a quay has overshot the edge of the quay and struck a crane, causing it to collapse. During berthing operations, rail-mounted cranes should be positioned where they cannot be struck by the ship.

10. Container cranes are designed primarily to lift containers and not people. Advice should be sought from the crane manufacturer if there is any doubt about precautions that should be taken when such cranes are used to lift portworkers, for example in a lashing cage. The precautions may include:

— reducing the operational lifting, lowering and trolleying speeds;

— prohibiting gantrying along the quay;

— use of "dead man's" controls in the operator's cab;

— double wire purchases;

— safety cut-outs;

— use of safety belts;

— limiting the number of persons to be carried at one time;

— fitting an emergency stop button accessible to those being carried;

— fitting a control monitoring system to detect serious errors;

— more frequent safety inspections.

5.2.4. Lift trucks

5.2.4.1. General requirements

1. An adjustable seat should be adjusted to provide a comfortable driving position. If a suspension seat is fitted, the weight adjustment should be set to correspond to the operator's weight to minimize transmission of shocks to the spine.

2. Appropriate fork attachments, such as rotating heads and drum or bale clamps, should be used whenever they are available to handle particular types of cargo.

3. A special attachment consisting of a frame fitted to the fork-anchoring frame and fitted with a conventional hook should be used only if:

— its safe working load, including that for traversing on a slope (if required), is marked on it;

— the maximum height of lift of the hook is conspicuously marked on the mast of the truck;

— care is taken to ensure that swing of the suspended load is controlled when the truck is travelling.

4. Trucks and battery containers of electric trucks that are hoisted aboard ship should be lifted by suitable slinging points.

5. Any oils spilled should be cleaned up as soon as possible.

5.2.4.2. Safe use

1. When lift trucks are used:

— flashing orange/amber lights should be operated whenever the truck is in motion;

— trucks should be driven at an appropriate safe speed; this should not exceed 25 km/hr;

— seat belts should be worn when appropriate;

— the clearance of the loads should be borne in mind, especially when the truck enters places that are narrow or restricted in height;

— stacking on and travelling across inclines should be avoided;

— if the load obscures forward vision, the truck should be driven in reverse or an observer used;

— the audible warning device (horn or klaxon) should be sounded when necessary to alert pedestrians and when the lift truck is about to pass through rubber swing doors or pass any concealed entrance, parked vehicle or large obstacle, such as temporary cargo;

— truck forks or other load attachments should be fully lowered when the truck is parked;

— the parking brake should be applied whenever the truck is at rest.

2. Lift trucks should not be:

— driven:

  • without permission;

  • on routes other than those which have been specifically approved in advance;

  • with unsafe loads;

— braked unnecessarily sharply, made to take bends at high speed, or otherwise driven dangerously;

— used to:

- lift a load exceeding the truck's capacity;
- lift a poorly balanced load;
- lift a load on only one fork arm;
- travel with the forks raised above a nominal 150 mm, either loaded or unloaded;
- carry persons on trucks not specially equipped for the purpose, on trailers with or without brakes, on couplings, or on the forks;
- pull or push a wagon or other vehicle with a truck not specially designed for the purpose, unless a special safe system of work has been drawn up by a competent person;
- deposit metal goods where they might fall onto the batteries of electric trucks;

— used with any additional weight on the counterweight;
— left in a traffic lane;
— left with the ignition key in an unattended truck.

3. Special care should be taken when a truck is driven:
— on slippery ground;
— in areas where there is loose dunnage or waste material;
— by or through doorways used by personnel;
— around corners where vision is restricted;
— in any place where the overhead clearance is limited;
— near any open hatch or lift opening on a ship when the lift platform is away from that deck;
— on bridges over trenches or other gaps.

4. During stacking and unstacking operations with a counterbalanced lift truck (figure 58 (1 and 2)):

Figure 58.   Stacking and unstacking by lift truck

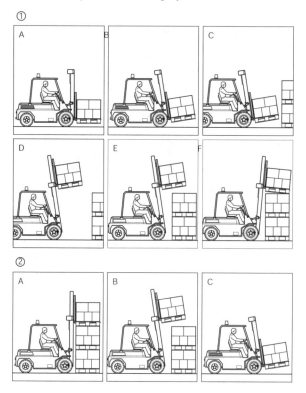

1. Stacking.

A. Take the load at ground level and raise about 150 mm. B. Give lifting assembly and load maximum tilt to rear. C. Align the truck in front of the stacking point and set the brakes. D. Lift the load to the required height. E. Move the truck forwards slowly until the load is aligned over its final position and reset the brakes. F. Lower the load slowly, if necessary allowing the platform to tilt forwards slightly.

2. Unstacking.

A. Pick up the load with the uprights vertical. B. Move backwards with load and tilt to rear. C. Lower the load.

— forks should penetrate under the load as far as the heel of the forks;

— forks should be at least three-quarters the length of the load in the direction of the forks;

— where loads are stacked behind one another, the fork length should be such that it does not disturb the stack behind the load being lifted;

— when travelling, with or without a load, the forks should be at least 150 mm above the ground, so that there will be no risk of the forks or load hitting the ground;

— no load should be carried or raised with the mast tilting forward, unless the truck complies with national or international standards relating to operations.

5. During stacking (figure 58.1):

— the stack should be approached slowly, with the mast tilted backwards;

— when the truck is sufficiently close to and facing the stack, the brakes of the truck should be applied and the forks then raised until they are slightly above the stacking level;

— when the load is over the stack, the brakes should be re-applied, the mast should be brought to its vertical position and the load deposited;

— once the load is properly stacked, the forks should be withdrawn from beneath the load (with the mast tilting forward if necessary) by backing the truck away from the stack;

— the forks should then be lowered to the travelling position.

6. During unstacking (figure 58.2):

— the truck should approach the stack and stop with the fork tips approximately 300 mm from the stack face;

— the operator should check that the forks are at the correct width spacing and the load is within the capacity of the truck;

— with the forks raised to the correct height and the mast vertical or tilted slightly forward, the truck should be moved forward until the heels of the forks are in contact with the load, and the brakes of the truck should then be applied;

— the forks should be raised until the load is just clear of the stack and the mast should be tilted slightly backwards. Great care should be taken to ensure that any other load on the stack is not disturbed during this operation;

— the operator should ensure that the way is clear, and should reverse the truck sufficiently far from the stack to clear the road;

— the load should then be lowered to the travelling position, the mast should be tilted fully backwards and the truck should then move off steadily.

7. When a counterbalanced truck is driven on an incline:

— the load should always face up the slope;

— without any load, the forks should face down the slope;

— travelling across the incline and turning on the incline should be avoided.

8. A counterbalanced truck should not pick up, put down or carry a load on a slope that runs across the fore-and-aft centre line of the truck.

9. When a truck is travelling on the platform of a ship's lift, it is essential to ensure that:

— no part of the truck or its load projects beyond the edge of the platform;

— the truck brakes are firmly applied;

— the operator stays at the controls of the truck.

10. Pallet loads should be secure and safely banded. They should not overhang the pallet.

5.2.4.3. Reach trucks

1. A reach truck (figure 59) should not be driven with its reach mechanism extended.

Figure 59. Reach truck

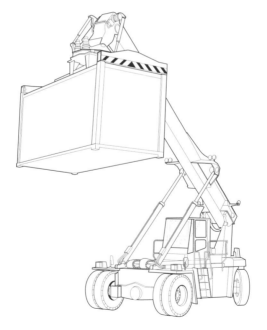

2. Before the reach mechanism is operated, the brakes of the truck should be properly applied.

3. No person should be allowed to step over the reach legs while the truck is in use.

4. A check should be made to ensure that the load is raised above the reach legs before they are retracted.

5.2.4.4. Side-loading forklift trucks

1. When using a side-loading forklift truck (figure 60), the load should be raised clear of the deck before traversing in.

2. If stabilizing jacks:

— are fitted, they should be fully lowered before the load is lifted;

— are fitted but not used, and the truck has a reduced safe working load when used without stabilizing jacks, this load should not be exceeded;

— are not fitted, the load should not exceed the load appropriate to operating without stabilizing jacks.

3. Unless backward tilt of the forks is used to stabilize a loose load, the load should be firmly on the deck and the forks just clear of the ground before travel takes place.

4. If the truck has a jackless capacity, it should not move while the mast is in the traversed-out position other than to manoeuvre the load into position, for example, on a vehicle platform.

5. When a side-loading forklift truck is used for stacking:

— the stack should be approached with the load placed on the deck of the truck, making use of backward deck tilt (if fitted);

Figure 60. Side-loading forklift truck

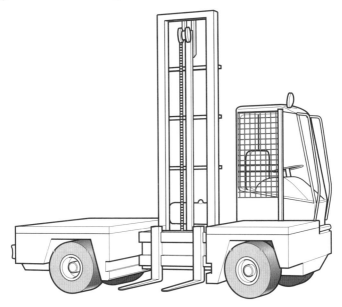

— the truck should stop when the load is in line with the depositing position and the truck is parallel to the stack;

— the stabilizing jacks, if any, should be firmly applied;

— any deck tilt should be removed;

— the load should be raised to the required height;

— the load should be traversed out until it is over the stacking position;

— the load should be lowered onto the stack, any tilt being corrected as necessary;

— when the load is properly stacked, the forks should be lowered until free of the pallet or dunnage strips;

— the mast should be traversed fully in and the forks lowered to just below deck level;

— the stabilizing jacks, if any, should then be retracted or raised.

6. The procedure for unstacking should be the reverse of the procedure for stacking.

5.2.4.5. Batteries

1. The batteries of a truck should be handled, whether for charging, removal or other purposes, only in a proper place especially set aside for that purpose and under the supervision of an experienced person.

2. Only authorized and competent persons should handle batteries owing to the possibility of injuries from electric shock or burns from battery acid.

5.2.4.6. Pedestrian-controlled pallet trucks

1. The operator of a pedestrian-controlled pallet truck should always walk with it and not attempt to ride upon it.

2. The operator should walk to one side of the control handle and clear of the truck, if it is necessary to precede the truck.

3. When approaching an obstacle, the operator should be behind the truck whenever possible.

4. When a truck is to be driven onto a vehicle being loaded or unloaded, it is essential to verify that:

— the vehicle's brakes are firmly applied;

— the bridge spanning the gap between the loading platform or bay and the vehicle is sound, of adequate strength and firmly positioned;
— the vehicle's loading surface is sufficiently strong and in good and level condition.

5. When a truck is required to use a goods lift, the operator should:
— approach the lift load first;
— stop at a safe distance from the gate;
— check that the combined weight of the truck and its load is within the safe working load of the lift;
— check that the floor of the lift is level with the ground or loading floor;
— check that the load will clear the lift entrance;
— drive on slowly and cautiously;
— firmly apply the brakes and shut off the power.

5.2.5. Other lifting appliances

1. Mobile elevating work platforms (MEWPs) should be used only on fully guarded platforms (figure 61). If they are to be used as a means of access, the manufacturer should be consulted about necessary precautions.

2. Particular attention should be paid to the stability of MEWPs. Before the platform is raised it is important to ensure that:
— the appliance is suitable for the intended operation;
— the ground below its wheels and outriggers can support the load;
— the outriggers are fully extended and, if necessary, supported by suitable packings;

207

Figure 61.   Mobile elevating work platform (MEWP)

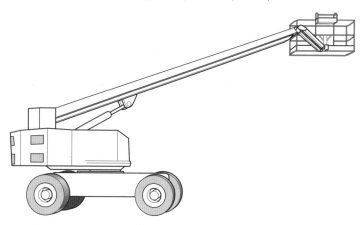

— wheel locks are applied, if fitted;

— the carriage is level.

3. MEWPS should only travel with the platform raised if they have been designed to do so. Travel should be at slow speed, special attention being paid to avoid potholes or slopes that could reduce stability.

4. Winches operated by steam power should be operated in such a way that:

— portworkers are not scalded by hot water or steam;

— exhaust steam does not obscure the operator's field of vision;

— the cylinders and steam pipes are cleared of water by opening appropriate drain cocks;

— a constant steam pressure is maintained at the winches to ensure safety and smooth working while the winches are in operation.

5.2.6. Use of more than one lifting appliance to lift a load

The use of two lifting appliances in tandem is a hazardous operation that should be performed only in exceptional circumstances. It calls for detailed planning and great caution. In particular:

— it should be directly supervised by a competent person;

— only identical lifting appliances should be used;

— the load should not exceed the safe working load of either appliance by more than 25 per cent;

— neither appliance should lift more than 75 per cent of its safe working load;

— movements should be slow and strictly controlled;

— only one motion should be used at a time;

— as far as possible, cranes should not slew with the load;

— side loading of cranes should be avoided.

5.3. Loose gear

5.3.1. Safe working load

1. The safe working load (SWL) of items of loose gear should be determined by a competent person. Usually this is done by applying a factor of safety to the breaking load of the item, but the safe working load of specially designed lifting beams, lifting frames and special clamps should be determined by design calculations.

2. The safe working load of a sling (figure 62) depends on the configuration in which it is used (mode factor).

Figure 62.   Hitches/mode factors

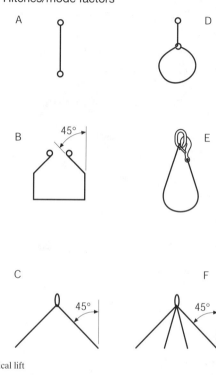

A. Straight vertical lift
   M = 1.0

B. Basket hitch 0°–45°
   M = 1.4

C. Two equal single legs used 0°–45°
   M = 1.4

D. Choke hitch
   M = 0.8

E. Single leg hooked back
   M = 1.0

F. Three or four equal single legs used 0°–45°
   M = 2.1

3. Using the uniform load method that is normally employed, the mode factors which should be applied to the safe working load of a single sling are:

Configuration	Mode factor
Straight vertical lift	1.0
Choke hitch	0.8
Vertical basket hitch	2.0
45° basket hitch	1.4

4. Using the trigonometric load method, the SWL of an inclined single sling can be calculated from the following formula:

$SWL = 1 \times SWL$ of single leg $\times \cos \alpha$

(where α is the angle of the sling from the vertical).

5. Using the uniform load method, the mode factors for multi-leg slings of wire, chain or fibre with a maximum angle of 45° from the vertical are:

Sling	Mode factor
Two leg	1.4
Three leg	2.1
Four leg	2.1

6. Using the trigonometric load method (figure 63), the SWL of a multi-leg sling can be calculated from the following formulae:

Two-leg sling – $SWL = 2 \times SWL$ of single leg $\times \cos \alpha$

Three- and four-leg slings – $SWL = 3 \times SWL$ of single leg $\times \cos \alpha$

(where α is the angle of the sling leg from the vertical). The SWL of a four-leg sling should be the same as that of a three-leg sling as most loads are not uniform.

211

Figure 63.   Examples of stress on a two-leg sling using the
            trigonometric method of calculation

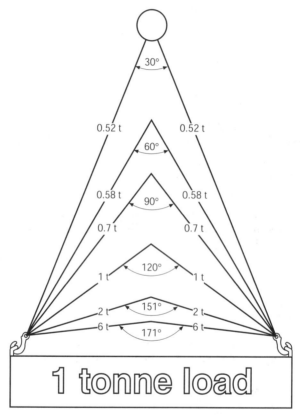

7. In normal use, the angle of 45° from the vertical should
not be exceeded. If this becomes necessary, the angle
should never exceed 60°, since at this angle the stress in
each leg of a two-leg sling is equal to the weight lifted.

5.3.2. Safe use

1. Slingers and other persons responsible for attaching loose gear to the load and lifting appliances should be:

— trained and competent in slinging and directing the movements of lifting appliances;

— capable of selecting the correct loose gear;

— able to recognize defects that should result in the rejection of gear;

— able to assess and balance loads;

— familiar with the signalling system in use in the port;

— able to initiate the movement of the lifting appliance.

2. If more than one slinger is required for a particular load, one slinger should be in charge of the operation and should be the only person to direct the lifting appliance operator.

3. Every item of loose gear should be visually inspected by a responsible person before use.

4. Any item of loose gear seen to be defective on inspection or during use should be taken out of service and referred to a competent person.

5. Loose gear should not be:

— dropped from a height;

— subjected to snatch or shock loads.

6. A sling should not be:

— used if crossed, twisted, kinked or knotted;

— used to roll a load over;

— dragged from beneath a load by a lifting appliance, unless the load is resting upon dunnage of adequate thickness;

— subjected to excessive heat or allowed to come into contact with any acid, alkali, abrasive or other substance liable to damage the sling.

7. Before the hoisting signal is given to the operator of a lifting appliance after a load has been released, it is essential to confirm that:

— the sling is completely free of the load;

— any hook or other lifting device at the end of the sling is hooked or attached to the upper ring of the sling; if that is not practicable, steps should be taken to ensure that the hook or other lifting device will not catch or foul any object.

8. A shackle should not be used on a sling unless it is fitted with a proper shackle pin; an ordinary bolt or piece of steel bar should not be used.

9. The links of a chain should not be joined together by a nut and bolt, by wiring, or by passing one link through another and inserting a bolt or nail to hold it in place.

10. A chain, fibre rope, wire rope or webbing sling should not be allowed to come into contact with any sharp or jagged edges of the load, but should be protected by means of wood, webbing, rubber or other suitable packing.

11. If a load has sharp angles, wedges of rag, paper, wood, plastic or rubber tyre should be placed over the edges so as not to damage the sling (figure 64).

12. No hook or other lifting device should be attached to any wire, strap, band or other fastening of a load unless it is so attached for the purpose of breaking out a load and

Figure 64.   Methods of protection of slings etc. against sharp edges

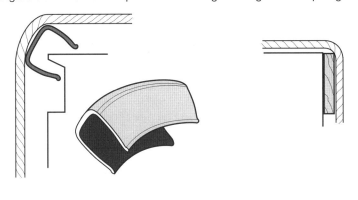

only lifted a short distance to make up a set. Unitized loads or packages banded by twisted wire or flat metal should only be lifted by such bands if they are accompanied by relevant certificates in the same way as disposable or one-trip slings.

13. In the case of unitized loads (figure 65), hooks or lifting devices may be so used, provided that:

— they are specifically designed for the purpose;

— the wire, strap, band or other fastening has been properly secured to the load;

— the wire, strap, band or other fastening is compatible with the hook or lifting device used;

— at least two hooks or lifting devices are used, and each is secured to a band or group of wires.

14. No hook should be attached to the rim or chine of a drum or barrel unless the hook is of suitable shape, and the rim or chine is of adequate strength and depth for the purpose and is not distorted or otherwise damaged.

15. No hook should be inserted into the attachment of a load unless the attachment is of sufficient size for the load to be freely supported on the seat of the hook. The load should never be applied to the point of the hook nor should the hook be hammered in.

16. When lifting a heavy or bulky load, care should be taken not to crowd the hook of the lifting appliance with slings.

17. If a large number of slings cannot be avoided, one or more bow shackles should be used to connect the slings to the hook.

18. When it is necessary to handle irregularly shaped loads, such as a machine tool or very long loads where the centre of gravity may be some distance from the vertical geometric centre line, a number of trial lifts should be made by partially lifting the load and adjusting the sling position until the suspended load is as level as practicable.

Figure 65.   Lifting banded unitized loads

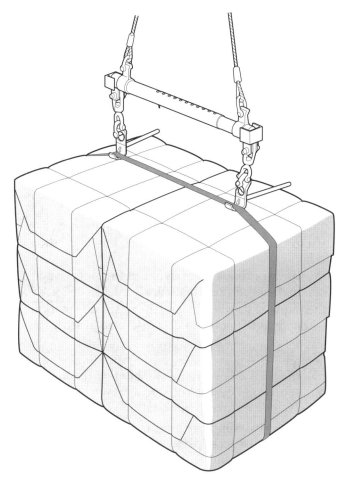

19. Where it is necessary to shorten one or two legs of a sling in order to achieve equal balance, a proper device such as a chain claw should be used (figure 66). A sling should never be shortened by knotting.

20. When tubes, girders, long metal sheets or similar long loads are lifted, the safest and most appropriate means should be employed.

21. Where two shoes, dogs or hooks used to grip a load are connected by a running chain, this should be fitted with a shortening clutch so that an angle of 60° can be maintained by the chain.

22. Where necessary, the load should be fitted with lanyards or guys (tag lines) so that twist or swing of the load can be controlled by persons stationed on the guys.

23. Unless a load is of sufficient length to warrant the use of a spreader beam or lifting frame, its weight should not exceed:

— the safe working load of either of the slings when slings of equal safe working loads are used;

— the rating of the sling having the lesser safe working load when slings of unequal safe working load are used.

24. The weight of a load to be lifted by a sling used in choke hitch (with the standing part of the sling being reeved through the hook or eye at the end holding the load) should be limited to 80 per cent of the safe working load marked on the sling (see section 5.3.1, paragraph 3).

25. When two slings are used, the slings should be passed around the load at least twice before being attached to the standing part of the sling (a wrapped choke hitch) in order to reduce to a minimum the tendency of the slings to slide inwards towards one another when they are under tension.

Figure 66.   Chain shortening claw

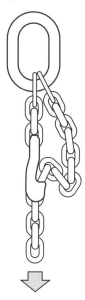

26. When a choke hitch is used, the angle between the hook or eye and the standing part should be allowed to take a natural position and not be knocked down to tighten it. A practical rule is to keep the height of the choke above the load to at least two-thirds the length of the sloping part of the sling. This rule is easy to apply to the slinging of sacks, but the load should always be well balanced.

27. When a chain sling is used in choke hitch, the hook or eye of the sling should be hooked or reeved into the standing part so that the subtended angle between the standing part and the end of the sling is not more than 90°. Slings other than

chain slings tend to take up an angle of 90°, but on chain slings this is prevented by the hook or eye locking itself between two chain links.

28. In the case of a sling having an eye at each end and reeved with both eyes on the hook of the lifting appliance and the two standing parts of the sling reeved through the eye of another sling placed around the load, the weight of the load to be lifted should not be greater than the safe working load marked on either of the slings.

29. Pre-slung slings are subject to all the normal requirements of manufacture and certification.

30. During the discharge, pre-slung slings should be inspected prior to each lift as damage can occur while the ship is at sea as a result of movement between the packages. Ideally, small dunnage sticks should be placed between the packages on loading to prevent chafing.

31. Ships carrying pre-slung cargo should hold a copy of the test certificate or certificate of conformity showing the safe working load of the slings and, if they are reusable, a copy of a current (i.e. issued within the last 12 months) thorough examination certificate.

32. Plate clamps should be of an adequate size and strength for the loads handled, and all the teeth on a clamp face and/or locking cam should be sound.

33. A self-locking plate clamp should not be used unless it is fitted with a safety catch to prevent the load from being accidentally released in the event of the tension upon the clamp becoming momentarily slack.

34. When a load is lifted by barrel hooks, crate clamps or similar appliances, the sling should be reeved from the

crane hook through the barrel hook, crate clamp or similar appliance and then back to the crane hook. In order that the resultant force will make the hook or clamp engage more firmly, an angle of 60º should be maintained between the legs of the sling.

35. Except when making up a set in circumstances in which portworkers could not be injured, the lifting of such loads as bales by the insertion of hooks should be prohibited.

36. Small loose goods such as small drums, canisters, boxes and carboys should be loaded onto suitable pallets or trays hoisted by four-legged slings. When necessary, special precautions, such as fitting a net around the slings, should be taken.

37. Buckets, tubs and similar appliances should:

— be loaded in such a way that there is no risk of any of the goods falling out;

— be secured to the hook by a shackle, unless fitted with a handle specially designed to fit the hook of a lifting appliance;

— have a handle with a special bend at its centre or be shaped in such a way that the hook or shackle will lift the bucket or tub only at the centre of the handle;

— have, if the handle can hinge about its attachments to the bucket or tub:

• hinge points above the centre of gravity of the bucket or tub when it is loaded; and

• a locking device fitted to prevent a bucket or tub from accidentally turning over when it is suspended.

38. When a cargo, such as loaded bags, sacks or reels of paper, is to be hoisted by a sling:

— an endless fibre rope or flat endless webbing sling should be used, and should be reeved in choke hitch in such a way that the two parts of the rope encircling the bags or sacks are spaced approximately one-third the length of bag away from each end;

— in the case of paper reels of large diameter, when three reels are hoisted at the same time by means of a sling, the reels should be placed in triangular fashion with one reel resting upon the other two;

— the bags or reels should be arranged so that their ends are all approximately in the same vertical plane.

39. When the hook of a multi-legged sling is attached to an eye fitting on a pallet, tray or load, it should be inserted into the eye from the inside of the load, so that in the event of a leg of the sling becoming momentarily slack, the hook will remain engaged in the eye (figure 67).

40. When ingots are hoisted, they should be supported by special bearers having eyes through which the slings are reeved in accordance with the guidance for lifting with barrel hooks, each layer of ingots being laid at right angles to the layer beneath, or by another suitable and safe method.

41. When a reel of cable or coils of metal wire are to be lifted, the slings should be attached to a steel bar of adequate strength and length passing through the hole in the centre of the reel, or through the coils of wire (figure 68). Such bars should be tested and certified in the same way as any other item of loose gear. Alternatively, a sling that has been specially designed for use with reels or coils may be used on its own.

Figure 67.   Correct positioning of lifting hooks

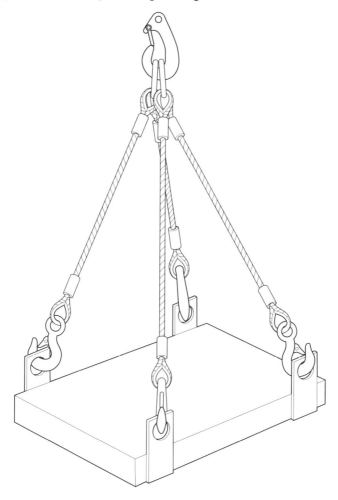

Figure 68.   Woven steel sling for lifting wire coils

42. Animals hoisted should be in boxes, cages or slings that immobilize them sufficiently to prevent dangerous disturbances of loading or unloading operations, or injury to portworkers or to the animals themselves.

5.3.3. Ropes and slings

5.3.3.1. Use

1. Grade "T" slings should never be exposed to acid or sulphur in the atmosphere, as hydrogen embrittlement may cause a critical loss of strength.

2. Chains intended for use for significant periods at temperatures below about –5°C should be made of special steels (see also section 4.4.2, paragraph 2).

224

3. Wire slings should not be used at temperatures above 100°C, as they may have fibre cores and/or alloy ferrules.

4. Natural fibre slings are often used for handling light cargo. The use of ropes treated with rot-proofing and/or water-repellent agents can reduce damage by damp.

5. Natural and synthetic fibre ropes and slings that become wet should be dried naturally.

6. A natural or synthetic fibre rope intended for use with a boatswain's chair should be suitably tested before a person is hoisted in the chair.

7. Ropes composed of synthetic fibre should not be surged, paid out or rendered by slacking away the rope, as this may subject it to frictional heat. They should also be protected from heat generated by other external sources. Any melting on the surface will render the rope or sling useless.

8. Polypropylene fibre ropes and slings should not be continuously exposed to ultraviolet light such as bright sunshine.

9. Nylon (polyamide) ropes and slings should not be immersed in water or wetted appreciably, as this can result in approximately 15 per cent loss of strength.

10. When not in actual use, synthetic fibre ropes and slings should be kept covered by tarpaulins, stowed below deck or in the store.

11. Disposable or one-trip slings should be scrapped by cutting up or other suitable means after being taken from their loads at the final destination.

12. The normal mode factors for slings do not apply to disposable or one-trip slings, as the safe working load at an

angle has already been calculated. However, if the safe working load is only given for a single angle (such as 0°), the normal mode factors should be applied.

5.3.3.2. Storage and maintenance

1. Loose gear (such as chains, wire rope and fibre ropes) when not in use should be stored under cover in clean, dry, well-ventilated places, free from excessive heat and protected against corrosion.

2. Loose gear should be raised from the ground and not be in contact with damaging agents such as ashes, clinker, coke breeze or chemicals.

3. As far as is practicable, loose gear in storage should be so arranged that items with the same safe working load are grouped together and fibre ropes separated from metal gear.

4. Ropes should be carefully wound on reels of wood, metal or plastic, or laid out in straight lines so as to avoid kinks and partial unravelling.

5. Synthetic fibre slings should be hung on wooden pegs or galvanized hooks away from any source of heat.

6. Ropes and slings that are wet should be dried naturally.

7. Ropes or slings that have been, or are suspected of having been, in contact with any acid, alkali, gypsum or other substance harmful to them should be destroyed.

8. If it is suspected that a synthetic fibre rope or sling has come into contact with organic solvents such as paint, paint stripper or coal tar, it should be thoroughly washed as soon as possible with fresh water, allowed to dry in air and be inspected for damage.

9. A synthetic fibre rope should not be respliced if worn. Flat woven webbing should not be repaired or altered.

10. Loose gear in stores should not expose workers to risks of overreaching, or of falling objects.

11. Loose gear awaiting repair should be clearly identified, recorded and stored separately in a quarantine area.

12. Loose gear beyond repair should be scrapped, or held in a clearly marked area and identified for scrapping by marking with an agreed colour, or by some other means.

5.3.3.3. Removal from service

1. When loose gear is inspected or examined, particular attention, as appropriate, should be paid to:

— illegible markings;
— broken, missing, distorted, worn, corroded or otherwise damaged components;
— chemical attack;
— heat damage;
— solar degradation.

2. Particular attention should be paid to the effects of cuts, chafing and damage to stitching of sythetic fibre slings.

3. Loose gear should be removed from service for scrapping if:

— wear in eyes of chain links or the saddle of hooks exceeds 8 per cent;
— permanent elongation in sling chains exceeds 5 per cent;
— the cross-section of a chain link is reduced by more than 12 per cent;
— jaw openings of hooks have increased by more than 10 per cent;

— the diameter of wire ropes is reduced to below 90 per cent;
— the number of broken wires or needles in any length of ten diameters exceeds 5 per cent of the total number of wires in a rope;
— broken wires:
  • appear in one strand only;
  • are concentrated in a shorter length of rope than ten diameters;
  • appear in the tucks of a splice;
— there is more than one broken wire immediately adjacent to a compressed metal ferrule or fitting.

5.3.4. Other loose gear

1. When pairs of shackles are selected for a job, both should have the same safe working load. Size may be misleading, as grade "T" shackles are approximately twice the strength of grade "M" shackles.

2. "Dee" shackles should be used for straight pull applications and "Bow" shackles where a number of slings pull at different angles. Where shackles are permanently rigged, the pins should be locked by mousing a screw collar pin or by a split cotter pin on a nut and bolt pin.

3. The safe working load of a shackle in a sling should always be equal to the sling, the increased stress due to an angle in the arrangement being duly taken into account.

4. When used in normal slinging applications, the screw pins of shackles should only be done up hand tight. However, the pins should be secured with seizing wire to keep them from coming undone.

5. Pulley blocks selected for use should always have sheaves matched to the fibre or wire rope to be used. The diameter of the sheaves at the bottom of the groove should not be less than:

— 14 times the diameter of a wire rope;

— 5.5 times the diameter of a fibre rope.

6. Unless the alignment of the sheaves of a pulley block is to remain in line with a fixed fitting, a swivel head block should always be used.

7. A pulley block should:

— be regularly lubricated;

— not have its data plate or any grease nipple painted over;

— be kept in the ship's cargo store or stevedores' store when not in use.

8. Hooks should be selected to cause as little distortion and damage to the eye of a sling as possible. The larger the hook that is used, the less distortion to the sling.

9. Hooks should always have a means of preventing a sling from becoming accidentally detached.

10. Hooks are designed to take loads vertically through the saddle. Bow shackles should be used when there are too many slings in a hook or the spread is too wide. Shackles should always be used with their pin in the hook.

11. Specialized cargo-handling hooks should be used where appropriate. These include hooks designed to lift by specially designed bands around cargo or to stick into goods such as logs and bales.

12. Where hooks are hooked into the eyes of lugs or container corner fittings, they should always be hooked from the inside out to prevent them from becoming unhooked accidentally.

13. The correct type of loose gear should be used to lift ISO containers without spreaders, when this is permitted by international standard ISO 3874, *Series 1 freight containers – Handling and securing*. Those for lifting from bottom corner fittings fit in from the side and can be used vertically or at an angle. As they are right- and left-handed, it is important to check carefully that they are at the correct corner.

14. When grabs are used to handle bulk cargo:

— there should be ample room at loading and unloading points for workers to avoid the swinging grab;

— grabs should be secured against accidental opening and be so constructed that they can be locked in the open position to prevent persons from being trapped by accidental closing;

— if heavy goods such as ore are being handled, special supervision should be provided for trimmers;

— the attachment and changing of grabs on the lifting appliance should be left to the engineers in charge of the appliance.

15. Automatic container spreaders should be used whenever practicable. If manually operated spreaders are used, portworkers usually have to go on top of containers to hook on and off (see section 7.8.3).

16. Manual spreaders should always be fitted or removed on the ship's deck or quayside where the hook of the

appliance can be lowered. "Tag" or restraining lines should be used to control the container when necessary.

17. Vacuum and magnetic lifting gear should:

— only be used in holds if portworkers are able to take shelter from any falling objects;

— never be used to transport persons.

18. Vacuum lifting gear should only be used on cargo specially wrapped for the purpose, or that has an otherwise suitable surface for vacuum lifting pads.

19. When vacuum lifting gear is used:

— each pad should support an equal part of the load, so far as is practicable;

— the load should be suspended horizontally, as far as possible;

— the surface of the cargo to be handled should be clear of any loose material that would prevent any vacuum pad from making effective contact with the surface;

— warning devices should be tested at the beginning of each week.

20. When magnetic lifting gear is used:

— the power to the magnet should not be switched on until the magnet has been lowered onto the load to be lifted;

— after the power has been switched on, the lifting motion should be delayed for a few seconds (up to ten seconds in the case of scrap metal);

— it should be carefully lowered on the load, not dropped;

— it should not be allowed to strike a solid obstacle;

— it should not be used to lift a steel sheet from a pile of sheets unless checks are made to ensure that sheets beneath the sheet to be lifted are detached;

— it should not be used on hot metal.

21. When magnetic gear is not in use:

— the power should be switched off to prevent the magnets becoming too hot;

— the magnet should be supported by suitable means; it should not remain on the ground.

22. Vehicles carried on non-purpose car carriers are slung either by means of special gear equipped with metal frames on which the chassis rests, or by fixing a net, usually a metal one, under the wheels, and attaching the net to ropes slung from a lifting beam (figure 69). It is essential to calculate the loads carried by each sling. The slings used should each be able to withstand the heaviest stresses that can be set up by a load.

23. The safest way of lifting pallets is with pallet forks having a sliding centre of gravity (figure 70). The tines of the forks should extend at least 75 per cent of the way under the pallet. These forks can be fitted with nets to prevent items falling from the pallet while in the air.

24. Other equipment for lifting pallets includes the following:

— *Spring-loaded pallet hoists*. These resemble a set of forklift tines hung from the crane hook. The spring-loading enables the centre of gravity of the appliance to adjust itself and keep the forks horizontal, whether loaded or light. They can also be adjusted for varying sizes of pallet.

Figure 69. Lifting vehicles by means of a sling

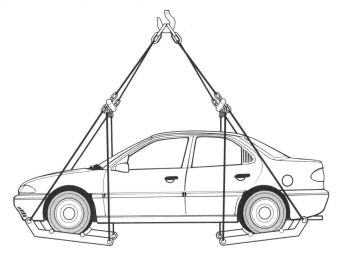

— *Pairs of metal stirrups*, each with a clamp or claw at each end. One end is fixed and the other can be extended by a spring. The stirrups engage under the edges of the pallet and grip the ends of the ties. A four-legged wire rope sling is attached to a pair of stirrups.

— *Wing pallets*. These should be at least 100 mm deep. A pallet bar should be placed under each wing and attached to a four-leg wire rope sling. When the pallet is lifted, two workers should stand by the set to ensure that the bars stay under the wings.

— Other appliances for lifting a factory pallet onto a shipping pallet to prevent the former from sliding on the latter. In their absence, special fastenings should be used to prevent sliding.

233

Figure 70.   Pallet lifting forks

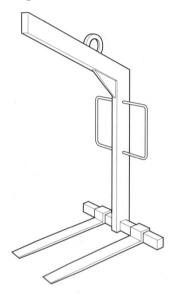

25. Pallets should never be lifted by slings passed between the boards, as it is likely that one will slip to the centre and allow the set to fall over. If the centre of gravity is too high in the set, a slight displacement of goods will allow the set to capsize.

26. Inspections of wooden pallets should include checks to ensure that:

— all deck boards are of equal thickness;

— all members are securely fastened by at least two nails that are adequately spaced;

— deck boards, bearers or blocks are not split or otherwise damaged or distorted;

— nails are not pulled through and do not project from deck boards;

— deck boards are not loose, permitting the pallet to distort or rack;

— members do not have extensive bark or knot inclusions;

— members are not contaminated by corrosive or flammable substances.

27. Pallets that are found to be defective should be destroyed or repaired before being returned to service.

5.4. Signallers

1. Signallers may be slingers or other persons responsible for giving directions to lifting appliance operators. They should be trained and certified in the art of signalling and directing crane movements by means of the signalling system in use in the port.

2. Only one person should act as the signaller for each lifting appliance. The signaller should be clearly identifiable to the operator and, unless responding to an emergency stop signal, the operator should only act on the signaller's instruction. Identification can be ensured by a distinctively coloured hat or clothing, or by radio call sign. Wearing light-coloured sleeves and gloves will enable signals to be more easily seen.

3. More than one signaller may be required for a lifting operation if:

— one signaller will not have a clear view of the load throughout its path of travel;

— hand signals are used and the first signaller has to move out of view of the appliance operator.

4. If signalling requires verbal communication, the signaller should be able to give clear and precise instructions in the language understood by the appliance operator.

5. Hand signals should be clear and precise, and given by wide movements that are unambiguous.

6. The system of hand signals should be agreed and clearly understood by all parties (figure 71). This is particularly important if the signaller and the operator of a lifting appliance are of different nationalities.

7. The signalling system should fail safe. If radios are used, each crane should have its own separate call sign and frequency, which should be kept free from communications for other purposes in order to prevent operators reacting to signals intended for another crane. The signaller should constantly repeat the required motion throughout the intended movement, such as "hoist, hoist... hoist", and the motion should be stopped if the operator ceases to hear the instruction.

8. The signalling system should include a means for a signaller to inform the crane driver that he/she will no longer be giving the directions. A further signal should indicate to the crane operator that a second signaller is taking over responsibility for directing the crane movements.

9. Signallers should not give an order before satisfying themselves that all measures have been taken to ensure that the operation can be carried out safely. The essential characteristics of signallers should be ceaseless vigilance and awareness that appliance operators are totally dependent on them during operations outside the operator's line of sight.

Figure 71. System of hand signals

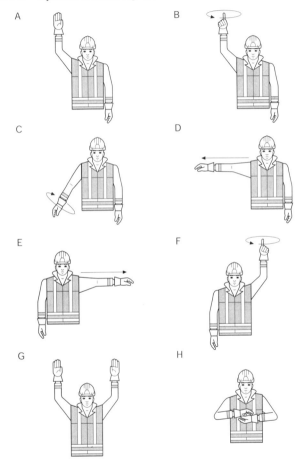

A. Stop (end of movement). B. Raise. C. Lower. D. Move in direction indicated. E. Move in direction indicated. F. Twistlocks on/off; rotate wrist of left hand. G. Emergency stop. H. End operations.

10. Before work is started for the day, a signaller should ensure that the workplace on the ship's deck or on the deck cargo is clear.

11. Signallers on ships should place themselves where they can be seen both by the workers in the hold and by the operator of the appliance (figure 72).

12. Signallers should do their utmost to protect persons against accidents. When necessary, they should warn persons in cargo holds, on lighters and ashore.

13. When cargo is being loaded or unloaded:

— by a fall at a hatchway, it should be possible for the signaller to pass safely between the hatchway and the ship's side;

— when more than one fall is being worked, a separate signaller should be used for each fall, except in the case of union purchase.

14. Before giving a signal to hoist, a signaller should ensure that the load is properly slung and that hoisting can be started without risk to persons working in the hold or elsewhere.

15. No signal to lower a load should be given by a signaller unless all persons are clear in the hold and elsewhere.

16. Before giving the signal to land, signallers should satisfy themselves that the load can be safely landed.

17. Signallers should never:

— give an order to move a load if any person is under its path; the person should be asked to move;

— agree to order operations that would violate safety rules, such as operations with defective slinging, dragging loads

Figure 72.   Correct positioning of signaller

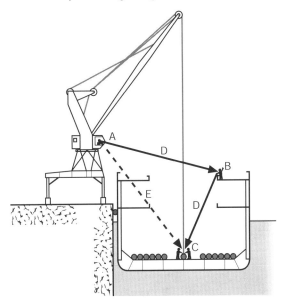

A. Crane driver. B. Signaller. C. Portworkers in hold. D. Direct line of sight. E. Direct line of sight impossible.

horizontally other than by bull roping, or with persons travelling on the load;

— give an instruction for operations if the light is insufficient or if there is thick fog, unless special precautions are taken.

18. Signallers should ensure that no persons are carried by lifting appliances except in properly constructed personnel carriers.

19. If it is necessary to stop a load while it is being raised or lowered, the signal should be precise but not abrupt, so that the operator of the lifting appliance does not jolt the load.

20. Equipment used for giving sound, colour or light signals for hoisting, lowering or transporting loads should be efficient, properly maintained and protected from accidental interference.

6. Operations on shore

6.1. General provisions

6.1.1. General requirements

1. Many cargo-handling operations that are carried out on shore are also carried out on ships. The guidance in this part of the code applies to all such operations. Guidance that is relevant only to operations carried out on ships is given in Chapter 7.

2. All port operations should be carried out in accordance with a safe system of work by portworkers who are appropriately trained and supervised. The safe system of work should enable a worker to stop an operation immediately when there is a risk to safety or health.

3. All plant and equipment used in port operations should be:
— of good design and construction;
— of adequate strength for the purpose for which it will be used;
— of sound material and free from obvious defects;
— inspected at appropriate intervals;
— properly maintained in a safe and efficient condition.

4. Routine fire inspections should be carried out. These should include inspections during periods when work is not in progress, as many fires result from smouldering and can break out several hours after their initial cause.

5. All means of escape in case of fire should be kept free from obstruction at all times. Flammable materials should never be kept under stairways.

6. There should be a clear policy on smoking. Smoking should be prohibited throughout the port area and on ships, except in designated areas. Smoking and no-smoking areas should be clearly identified.

6.1.2. Access arrangements

1. Safe means of access should be provided to all places where persons have to work.

2. Persons on foot should be separated from vehicles, whenever this is practicable.

3. Pedestrian walkways should not be used for other purposes.

4. Where access is needed through areas from which pedestrians are excluded, arrangements should be made for them to travel in a suitable vehicle. Access by crew, pilots and other visitors to ships at a container terminal could be one example. Persons on ships should be informed, by means of a gangway notice or otherwise, how to summon the transport.

5. Mobile access equipment, such as mobile elevating work platforms (MEWPs or "cherry pickers") should generally be used in preference to portable ladders. However, such equipment should always be used in accordance with the manufacturer's instructions, particularly those relating to the locking or scotching wheels and the use of outriggers.

6. When it is necessary to use a portable ladder (figure 73):

— the top of the ladder should rise at least 1 m above the landing place or the highest point to be reached by a person using the ladder, unless other adequate handholds are provided;

Figure 73. Use of portable ladders

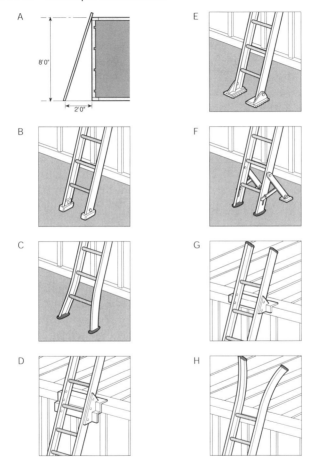

A. Correct angle of use. B. Rubber feet. C. Splayed stiles at foot of ladder. D. Double step angle piece. E. Safety feet. F. Stabilizing legs at foot of ladder. G. Angle locating piece at top of ladder. H. Splayed stiles that can be worked through at top of ladder.

243

— the stiles of the ladder should stand on a firm and level footing. Loose packing should not be placed under a stile;

— the ladder should be secured to prevent it from slipping. This should preferably be done by securing it at its upper resting place. If this is not practicable, it should be secured at its base. If even this is not practicable, the ladder should be footed by another worker;

— a ladder more than 6 m in length should also be secured at a point about one-third of its length from the ground;

— the ratio of the height of the ladder and the distance of its foot from the vertical surface against which it rests should be 4:1, i.e. 4 metres height, 1 metre out.

7. Workers using a ladder should:

— have both hands free for climbing up and down;

— face the ladder when climbing up and down;

— wear suitable footwear that is not likely to slip;

— use a belt or other suitable means to carry any object that is necessary.

8. A portable metal ladder or other mobile access equipment should not be used in any place where any part of it or a person on it is liable to come into contact with an overhead electric cable, trolley wires or any other electrical equipment with bare conductors, unless the power has been switched off and the system isolated. This should generally be in accordance with a permit to work that ensures that power cannot be switched on during the work.

6.1.3. Housekeeping and cleanliness

1. All parts of port areas should be kept in a clean and orderly condition.

2. All access routes and working areas should be kept free from objects and materials that are liable to cause a person to trip or slip.

3. Loose gear, tools and similar equipment should be kept safely or removed from working areas when not in use.

4. All dunnage and other rubbish should be collected as soon as possible and disposed of in an appropriate manner.

5. Spillages of oil or other materials likely to be a hazard should be cleaned up by trained personnel as soon as possible and reported to a supervisor.

6. Appropriate arrangements for clearance and gritting should be made to deal with snow and ice when necessary. Particular attention should be paid to means of access to workplaces, including access to ships.

7. All plant and equipment should be parked in appropriate designated areas when not in use.

6.1.4. Manual handling

1. Manual handling includes all forms of lifting, lowering, pulling and pushing of loads by portworkers.

2. Portworkers should not be required or permitted to manually handle loads that are likely to prejudice their health or safety owing to their weight, size or shape.

3. The need to handle significant loads manually should be avoided by the use of mechanical handling equipment, whenever this is practicable.

4. Manual handling should only be carried out by portworkers who have been trained or instructed in manual handling techniques in accordance with good kinetic handling

principles (figure 74). Supervisors should ensure that the correct lifting techniques are used in practice.

5. It is neither practicable nor desirable to prescribe the maximum weight that may be handled manually by a port-worker. Factors that should be considered include the weight of the load, the age, physique, posture, fitness and sex of the worker, the size and shape of the load, the working environment, and the frequency and duration of operations.

6. Particular consideration should be given to the loads that may be handled safely by workers under the age of 18 years and pregnant women. The employment of such persons may be restricted by national legal requirements.

Figure 74.   Manual lifting

7. Where appropriate, portworkers should be medically examined for fitness before being regularly assigned to carry out manual handling of significant loads.

8. Loads to be handled manually should whenever possible be compact. They should be clearly marked with their weight and provided with handles or other devices, as necessary. Handling aids should be used when appropriate.

9. The packaging of loads to be handled manually should not be liable to cause injury to persons handling them.

10. Appropriate personal protective equipment, including safety footwear and gloves, should be worn by portworkers engaged in manual handling.

6.1.5. Cargo in transit

1. Most cargo is only "kept" at ports for a short time while being trans-shipped or in transit along the transport chain.

2. Other cargo may be "stored" at ports for longer periods until it is needed.

3. All cargo in transit in a port should be kept or stored safely and securely.

4. Particular attention should be paid to the segregation of dangerous goods (see Chapter 8) and the need to maintain clear access for the emergency services in the event of a fire or other incident.

5. It should be possible to identify the nature, quantity and location of all cargo present in a port at all times. This may be done by electronic or other means. Areas containing dangerous cargo should be clearly identified.

6. Portworkers should be made aware of the general nature of the hazards of any cargo that they handle and of the precautions to be taken when handling specific cargoes.

6.1.6. Operational maintenance

1. All plant and equipment in ports should be regularly maintained in a safe and efficient condition, in accordance with the manufacturer's or supplier's recommendations, relevant national legal requirements and operational experience. This should be done on a planned, preventive basis, and should include periodic inspections and examinations, as well as physical maintenance.

2. Inspection and maintenance should be carried out on emergency equipment and personal protective equipment, as well as operational plant and equipment.

3. Maintenance and inspections should be carried out by engineering personnel, operators or users, as appropriate.

4. All persons carrying out maintenance and inspection duties should be trained in the relevant procedures and the identification of potential defects that may be found.

5. Safe means of access should be provided to all places to which maintenance personnel have to go. This should normally be permanent access.

6. All plant should be isolated before maintenance work is started. The isolation system should include lock-off facilities if the plant can be started remotely. Where necessary, maintenance work should be carried out in accordance with a "permit to work" system.

6.1.7. Hot work

1. Hot work should be carried out in accordance with national legal requirements and any port by-laws. These often require the permission of the port authority to be obtained before hot work is carried out. Obtaining a hot work permit from a port authority does not absolve those carrying out hot work from their duty to ensure that appropriate precautions are taken.

2. A hot work permit should generally specify:
— the location and nature of the work;
— the proposed time and duration of the work;
— the period for which the permit is valid;
— any precautions to be taken before, during and after the work;
— the person in direct control of the work;
— the identity of the person authorizing the work.

3. The precautions should generally include ensuring that:
— the work area is free from any flammable materials or flammable material residues. This should include the far side and adjacent areas of any plates or other metal involved in the hot work, and any area where hot particles may fall;
— no flammable or otherwise dangerous substances will enter the area during the work;
— the atmosphere in the work area is safe to breathe and remains so throughout the work;
— appropriate personal protective equipment, including overalls, gloves and eye protection, is used;

— appropriate fire-fighting equipment is available at the site of the work, together with a person trained in its use; this may be a person carrying out the work;

— any cylinders of flammable gas and oxygen, and hoses and torches attached to them, are removed from any enclosed space when the work stops;

— periodic checks for smouldering are made after the work has finished. These should include adjacent spaces that may have been subjected to heat or falling residues, as fires due to smouldering often break out several hours after completion of work.

4. If hot work is carried out on sprinkler systems in warehouses or elsewhere, particular care should be taken to ensure that other adequate fire-fighting facilities are available during the period that the sprinkler system is inoperative. If it is not practicable to remove goods from the area below the work, they should be covered by non-flammable sheets to protect them from falling hot particles.

5. The permit should include a facility for "signing off" when the work is completed.

6.1.8. Use of personal protective equipment (PPE)

1. PPE should never be used as a substitute for eliminating or controlling a hazard at source. However, if this is not possible, appropriate PPE should be provided and used.

2. PPE should be provided by the employer at no cost to the portworker and in accordance with the provisions of Articles 16(3), 17 and 21 of the Occupational Safety and Health Convention, 1981 (No. 155).

3. PPE should generally be available in a range of sizes, as one size or type seldom fits all. Comfort and acceptability

to the wearer are important, as the equipment may need to be worn for long periods.

4. The particular PPE that is necessary should be determined by an assessment of the hazards involved.

5. Portworkers should be instructed in the correct use and care of the PPE provided to them. They should use the equipment when required and take good care of it.

6. Managers and supervisors should ensure that appropriate PPE is used by all portworkers in accordance with instructions. Managers should give a clear lead by using the equipment when it is required.

7. In general, all portworkers should be provided with safety footwear, safety helmets, gloves and overalls, and should wear them when appropriate. Other types of PPE should be provided and worn as necessary.

8. All persons in cargo-handling areas should wear high-visibility overalls or other high-visibility outer clothing.

9. Loose clothing should never be worn by workers when working near open conveyors or other moving machinery. One-piece overalls are suitable.

10. Different-coloured overalls or other outer clothing and safety helmets may be useful to identify persons such as trainees and visitors. This helps to identify and draw attention to unauthorized persons in working areas or to persons who may be less familiar with hazards in the area.

11. Portworkers handling substances that are corrosive or can be absorbed through the skin should wear appropriate impervious personal protective clothing.

12. Portworkers who normally wear spectacles should not wear spectacles with glass lenses at work. Plastic lenses

are far less likely to cause injury in the event of an accident. Proper safety spectacles are preferable even in areas where they are not specifically required.

13. PPE that is not in use should be kept in suitable facilities. If the equipment or clothing may be contaminated by toxic or otherwise dangerous substances, it should be kept separate from the accommodation for workers' other clothing (see section 10.4). The facilities should be kept in a clean and orderly condition.

14. All PPE should be regularly cleaned and maintained in an efficient and hygienic condition, and replaced when necessary. This may be done centrally, or by the users if they have been appropriately trained. In all cases, maintenance should be carried out in accordance with the manufacturer's recommendations. Specialist equipment, such as some types of life jackets, may need to be returned to the manufacturer for periodic servicing.

15. Filters in respiratory protective equipment and other components with a limited capacity or shelf life should be regularly replaced in accordance with the manufacturer's recommendations.

16. Reusable PPE should be washed and disinfected, as appropriate, before being reissued.

6.2. Cargo packaging

1. Factors that should be considered when choosing packaging for cargo include:
— properties, including the weight, of the cargo;
— properties of the packaging;

— proposed method of stowage in a hold or cargo transport unit;
— climatic conditions to which the cargo will be exposed along the transport chain;
— legal requirements in countries along the transport chain.

2. Packagings and packages should be marked with relevant information as necessary. This may include:

— identification of the centre of gravity;
— identification of slinging points;
— nature of the cargo, such as "Fragile";
— correct orientation;
— dangerous goods labels, placards, marks and signs.

3. Traditional wooden casks, boxes and crates should not have projecting fastenings or sharp edges on metal reinforcements.

4. Wooden packaging and dunnage should comply with legal requirements relating to the importation of forest products that are aimed at the prevention of infestation.

5. Cardboard boxes or other cardboard packaging should generally not be used in very humid countries, as moisture can lead to crushing of the packages and cause stacks to collapse. Any signs of dampness on a cardboard package should be investigated and appropriate action taken. Dampness on packages may be caused by leakage from one or more receptacles inside them.

6. Paper bags or sacks should not be used in circumstances in which they are likely to be affected by atmospheric humidity or be exposed to strong sunlight for prolonged periods, as these can lead to deterioration.

7. As the properties of plastics vary widely, any plastic packagings selected should be appropriate for the intended cargo, journey and destination. Plastics are waterproof, and considerably stronger and lighter than many traditional packaging materials. They are generally suitable for use at temperatures between –15°C and +50°C, although some soften with heat and may degrade in ultraviolet light and prolonged strong sunlight. Plastic packages should be secured when necessary, as plastics have a low coefficient of friction and may be easily displaced.

8. Single-trip flexible intermediate bulk containers should never be reused.

9. All containers used in international transport, except offshore containers and those specifically designed for transport by air, should comply with the IMO International Convention for Safe Containers (CSC), 1972. There is no exemption for single one-way journeys.

10. The main requirements of the Convention are that all containers should:

— be of a design that has been approved by the administration of a contracting State following satisfactory testing;

— have a valid safety approval plate permanently affixed in a readily visible place on each container, normally on a door;

— be maintained in accordance with a periodic or continuous examination programme that has been approved by the administration of the relevant contracting State.

11. A periodic examination programme requires the container to be thoroughly examined in accordance with the programme within five years of the date of its manufac-

ture, and thereafter within 30 months of the date of the last examination. The date of each examination should be clearly marked on the safety approval plate.

12. An approved continuous examination programme (ACEP) requires the container to be thoroughly examined in connection with all major repairs or refurbishments and each on-hire/off-hire interchange. The interval between such examinations may not exceed 30 months. The dates of examinations are not marked on the safety approval plates of containers subject to an ACEP programme. Instead, they should carry a decal marked with the letters ACEP and the reference of the approved examination programme. The colour of the decal indicates the year of the last thorough examination of the container.

13. The presence of a valid safety approval plate on a container should not be taken to indicate that the container is in a safe condition. The plate can only reflect the condition at the time of the last examination. Damage or deterioration of the container may have occurred since that date.

14. Before goods are packed into a container, it is important to verify that the container has no obvious defects and carries a valid safety approval plate.

15. Offshore containers, defined as "portable units specially designed for repeated use in the transport of goods or equipment to, from or between fixed and/or floating offshore installations and ships", should conform to the guidance contained in IMO MSC/Circ. 860, *Guidelines for the approval of offshore containers handled in open seas*. The approval plate on an offshore container should be clearly marked "Offshore container".

16. The ISO standards for Series 1 freight containers are complementary to the CSC: ISO 830 *Freight containers – Vocabulary* defines the different types of container; ISO 668 *Series 1 freight containers – Classification, dimensions and ratings* specifies their designations, dimensions and ratings; and ISO 1496 *Series 1 freight containers – Specification and testing* details their specification and testing.

6.3. Container operations

6.3.1. Control of container operations

6.3.1.1. General requirements

1. For definitions of terms that relate to the handling of containers, see section 3.8.1.

2. The access of vehicles and pedestrians into container-handling areas should be strictly controlled.

3. No passengers in visiting container vehicles should be permitted to enter a container-handling area. Passengers should await the return of the vehicle from the container-handling area in a suitable waiting room.

4. All persons permitted to enter a container-handling area should be informed of the procedures they should follow while they are in that area. This may be done by signs, or by providing leaflets or copies of the relevant terminal procedures which they should follow. Different instructions will be relevant to different groups of people, such as terminal workers, drivers of visiting container vehicles, drivers of taxis and private vehicles, pedestrians and the crews of ships at berths in the terminal.

5. Instructions to drivers of container vehicles should specify where and when twistlocks securing containers to vehicles should be released or locked.

6. All containers arriving at a terminal by road or rail should be inspected for damage or tampering that could affect their safe handling. If a container is found to be unsafe, appropriate action should be taken.

7. The gross weight of all loaded containers should be known before they are lifted. Containers exceeding the maximum allowable weight of the container or the capacity of the relevant container-handling equipment should not be lifted.

8. The number of road vehicles permitted to enter straddle carrier and rail-mounted gantry crane (RMG) or rubber-tyred gantry crane (RTG) exchange areas at one time should be limited in order to reduce congestion.

9. Access to container-handling areas by pedestrians should be prohibited so far as is practicable. Any access that is permitted should be restricted to clearly designated walkways or under specific supervision.

10. No taxis or private cars should be permitted to enter container-stacking areas. Any taxis or private cars that are permitted to enter quayside areas should be required to keep to specified vehicle routes. They should not be permitted to enter a quayside area while containers are being loaded or unloaded from a ship. Where appropriate, a minibus or other suitable terminal vehicle should be provided to carry ships' visitors, ships' crews and other persons engaged in operations to or from such areas.

11. Vehicles from outside the terminal that need to leave the specified vehicle route should be escorted by a terminal vehicle.

12. All container terminal vehicles should be fitted with a flashing yellow warning light.

13. Containers should only be moved within the container terminal on vehicles that are suitable for the purpose.

14. All vehicles that have to be driven in a container-handling area while they are carrying containers that are not secured to them should be driven at an appropriate slow speed. Care should be taken to avoid heavy braking and sharp cornering.

6.3.1.2. Straddle carrier exchange operations

1. A straddle carrier exchange grid should only be used for loading and unloading freight containers from road vehicles. Grids should not be used as general waiting areas for road vehicles. Vehicles that need to wait for significant periods should be redirected to appropriate parking facilities.

2. Wherever possible, the grids should be operated with a one-way flow of traffic for road vehicles.

3. Where it is necessary for a road vehicle to reverse into a slot on a grid, there should be ample space for the manoeuvre to be carried out safely. Straddle carriers should only approach the slot from the opposite direction.

4. Reversing movements by road vehicles should not be permitted for any other purpose.

5. The entry of road vehicles to grid slots for loading or unloading should be controlled so that only one vehicle is in a slot at any one time.

6. Twistlocks and other container-securing devices should be released and locked in a designated safe place that should be clear of straddle carrier exchange grids.

7. The road vehicle driver should leave the cab of the vehicle and stand in a clearly marked area before the approach of a straddle carrier (see section 3.8.6, paragraph 4). This area should be forward of the cab of the vehicle, a safe distance from the vehicle and visible to the straddle carrier operator. The road vehicle driver should remain in the marked area throughout the loading or unloading operation. The driver should not return to the cab until the straddle carrier has left the grid.

8. A straddle carrier should only approach a road vehicle in order to load or unload it from the rear of the vehicle and should also leave the vehicle to its rear.

9. Any oversize container or problem container that cannot be handled safely at the grid should be moved to a suitable designated area where it can be dealt with safely.

6.3.1.3. RMG and RTG exchange operations

1. Twistlocks securing a container to a road vehicle should only be released or locked in a designated safe place. Where practicable, this should be separate from the place where the vehicle is loaded or unloaded. Particular care should be taken to ensure that all twistlocks securing a container that is to be lifted are fully disengaged.

2. Drivers of road vehicles should not stop on the marked runways of RMGs or RTGs.

3. Drivers of road vehicles should remain in the cabs of their vehicles at all times when in an RMG- or RTG-operated container-stacking area, unless specifically instructed otherwise.

4. When drivers of road vehicles have to be instructed to leave the cabs of their vehicles while in an RMG- or RTG-operated container-stacking area, they should only do so in accordance with a safe system of work. It is essential that where there is an overhead hazard the driver always wears a safety helmet and high-visibility clothing, and is visible to the RMG or RTG operator at all such times.

5. The driver of a road vehicle who needs to enter the ground cab of an RMG or RTG should only do so while the gantry is stationary. No more than one driver should be in the cab at any one time.

6. RMG and RTG operators should ensure that a road vehicle at which they are to work is stationary and that the cab of the vehicle is not under the intended lift.

7. Whenever practicable, containers should be transferred to or from a road vehicle to the side of the vehicle and not from the rear.

8. Visual and audible warnings of the movement of RMGs and RTGs should be given. Particular care should be taken if it is necessary to carry out "blind side" container exchange operations at the opposite end of the gantry to the operator's cab. Consideration should be given to the use of closed-circuit television (CCTV) systems or proximity alarms.

6.3.1.4. Entry to stacking areas

1. Entry into container-stacking areas should be restricted to authorized terminal vehicles that are fitted with flashing yellow lights and to authorized road container vehicles in RMG and RTG stacking areas.

2. No person should be allowed to enter a container-stacking area on foot other than by a clearly delineated walkway that does not cross a container traffic route. If a crossing of such a traffic route is found to be necessary, the crossing should be clearly marked and signed.

3. Work on foot in a container-stacking area should be done only with the express permission of Control. Permission should be given only after the relevant area has been isolated and Control has issued a permit to enter, and only for work that takes into account the characteristics of the terminal and the work to be carried out. A visual signal, such as a token or light on the control desk, should be used to remind the controller that an area has been temporarily isolated.

4. The permit to enter and to work in a specified area should be issued only by an authorized control officer who is satisfied that:

— all drivers of vehicles and plant operators in the container-handling area have been notified of the closure of the relevant area and have acknowledged the instruction. Each vehicle should have a board in the cab upon which the driver should enter a clear indication of the block that has been isolated;

— the relevant area has been clearly signed to prevent entry by other vehicles;

— the person or persons to enter the area are wearing high-visibility clothing in accordance with terminal rules;

— the person to enter, or the person in charge of the group if there is more than one person, has been issued with a two-way radio and is familiar with its use;

— the person or persons to enter the area have been fully instructed on the operations to be carried out, the procedures to be followed and the precautions to be taken.

5. The area to be isolated should include a suitable buffer zone between the location of the work and any active area. In a container-stacking area that is operated by straddle carriers, there should be at least one clear lane between the lane in which work is to be carried out and any active lane.

6. The particulars to be included in the permit to work should include:

— the name(s) of the person(s) wishing to enter;

— the work to be performed;

— time of entry;

— any specific instructions;

— an instruction not to leave the area until Control has been notified by radio and permission to do so has been received.

7. Control should ensure that any additional operators who enter the container-handling area, or relieve operators already in it at change of shift, are informed of the position of the isolated block.

8. The permit to work should be returned to Control once the work has been completed and permission to return has been received.

9. The permit should not be transferred if the work is not completed at the end of a shift or for other reasons, and

needs to be continued by others. A new permit should be issued in such cases.

10. Control should check the return of the permit for cancellation. If it has not been returned after a reasonable time, steps should be taken to find out why and, if necessary, the whereabouts of the missing person(s).

6.3.1.5. Emergency procedures

1. In an emergency, such as an accident or fire, Control should send a clear signal or instruction by radio or some other immediately recognized means. On hearing the emergency signal or instruction, all vehicles should immediately stop in a safe manner and remain stationary until instructed to do otherwise. When stopping, drivers and operators should bear in mind the need to leave clear access for emergency services and other rescue personnel and equipment.

2. Whenever possible, unserviceable vehicles, plant and equipment should not be repaired in an operational container-handling area. If it is necessary to do so, for example to enable them to be removed, the area should be isolated. This is particularly important in automated container terminals where unmanned equipment operates.

3. Unserviceable vehicles, plant and equipment should be clearly and appropriately marked to ensure that they are not used.

6.3.2. Container-stacking areas

1. The ground of all container-stacking areas should be maintained in a sound and level condition.

2. Every slot in a container-stacking area should be able to be readily identified. This may be done by the identification of blocks and rows on the ground or by other markings.

3. The tracks of RMGs and RTGs that service blocks of containers should be clearly marked and kept clear at all times.

4. Dangerous goods should only be kept in the stacking area in accordance with national legal requirements and terminal rules (see Chapter 8).

5. Containers in stacks should all be of the same length to ensure that the lower corner fittings of a container above the bottom tier rest directly on the top corner fittings of the container below. Non-standard-length containers may be stacked on standard containers, provided that their corner fittings are located in the same positions.

6. Containers should never be stacked beneath or close to overhead power lines.

7. Containers should not be stacked more than one high within 6 m of a building where there is a risk to persons in the building if a container is mishandled or subjected to high winds.

8. Consideration should be given to the possible effects of high winds on container stacks. This may include the orientation of containers in line with prevailing winds. Where necessary, containers should be secured by twist-locks or otherwise.

9. The ends of rows in stacks serviced by straddle carriers should be stepped down, where this is practicable, in order to improve the visibility of straddle carriers emerging from the stack.

10. Whenever practicable, tank containers should only be stacked one high. When it is necessary to stack tank containers more than one high, it is recommended that stacking cones be used, in view of the differences of tank container

frame designs. Tank containers carrying highly volatile substances should not be stacked above the pressure relief
valves of highly volatile flammable substances.

11. Any person seen on foot in a container-stacking
area, other than in an area that has been isolated, should be
reported to Control immediately. Control should isolate the
area until the pedestrian has been removed.

12. A conspicuous metal plate with a long handle may
be inserted into a top corner fitting of a refrigerated container connected to the electrical supply in order to prevent
it from being lifted while still connected (figure 75).

Figure 75.   Plate to prevent lifting of a refrigerated container while it is
connected to the electrical supply on the terminal

6.3.3. Container handling and lifting

1. Containers exceeding the maximum gross weight on their safety approval plate or the capacity of the handling equipment should not be handled.

2. Containers should be handled and lifted in accordance with relevant international standards. Table 1 of ISO 3874 *Series 1 freight containers – Handling and securing* illustrates the nine specified methods of lifting (figure 76). It should be noted that all methods have their limitations and many are not allowed for specified loaded containers.

3. Loaded containers should generally be lifted by container cranes vertically from their four top corner fittings with the aid of a purpose-designed spreader.

4. Empty containers may be lifted with the aid of a four-legged sling assembly (figure 77). The assembly may incorporate a chandelier spreader. The sling legs should be long enough to give a safe angle of not more than 90° between them at the crane hook. This angle should never be exceeded. Sling hooks inserted into corner castings should face outwards.

5. Containers carrying over-height loads may be lifted from the bottom corner fittings (figure 78) or with the aid of special purpose-designed over-height frames (figure 79).

6. The design of spreaders for twin lift operations should take into account the potential total gross weight of the two containers and possible asymmetrical loading of cargo inside them.

7. Containers should only be handled by other methods after careful evaluation of the equipment to be used and the methods of operation proposed.

Figure 76.   Summary of specified lifting methods: ISO 3874, table 1

Subclause	Description	Illustration
6.2	Top lift spreader	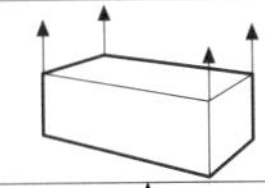
6.3	Top lift sling	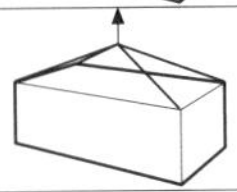
6.4	Bottom lift sling	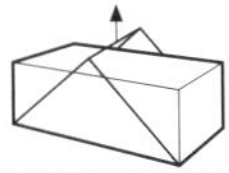
6.5	Side lift: method 1	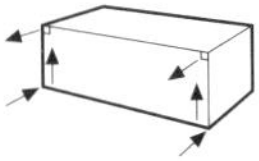
6.6	Side lift: method 2	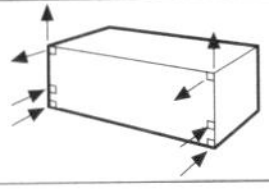
6.7	Side lift: method 3	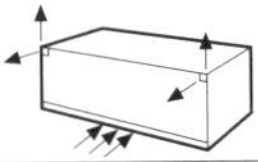
6.8	End lift: method 1	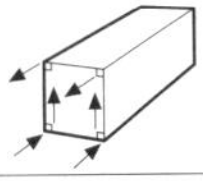
6.9	End lift: method 2	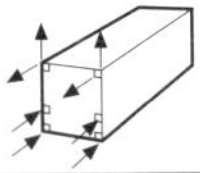
6.10	Fork-lift	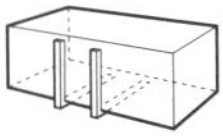

Source: The terms and definitions taken from ISO 3874:1997 *Series 1 Freight containers – Handling and securing*, table 1 (English version) are reproduced with the permission of the International Organization for Standardization (ISO). This standard can be obtained from any ISO member and from the website of the ISO Central Secretariat at the following address: www.iso.org (copyright remains with ISO).

Figure 77.   Top lifting of empty containers with a four-legged sling
            assembly

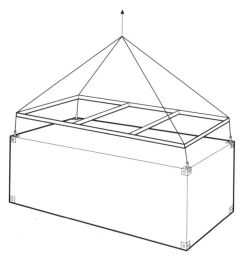

Figure 78.   Bottom lifting of containers

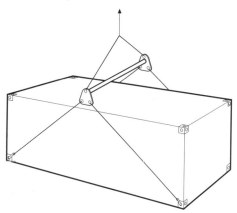

Figure 79. Over-height frame

8. Containers should only be handled by forklifts or goosenecks if they are fitted with forklift pockets or gooseneck tunnels, in accordance with ISO 1496, and provided that these are maintained in good condition. Tank containers should never be handled by forklifts.

9. Container-handling equipment should be driven at an appropriate safe speed. Speed should be reduced for cornering.

10. In order to maximize its stability, container-handling plant travelling with a container should carry it as low as is practicable to ensure adequate clearance of obstacles.

11. RMGs or RTGs lifting a container from a road vehicle whose operator is required to stay in the cab should lift the container slowly until it is seen to be clear of the vehicle.

12. Special precautions should be taken when it is necessary to handle damaged containers. Damaged containers should be withdrawn from service unless they are safe to continue to their destination for unloading or to a repair depot.

13. Hatch covers that are landed during loading or unloading operations should be clearly visible and not obstruct traffic routes. All relevant traffic vehicles and personnel should be alerted when hatch covers are landed.

14. The insertion of twistlocks into, or removal of twistlocks from, corner fittings of containers on the quayside should be carried out in accordance with a safe system of work that protects workers from the hazards of container-handling vehicles (figure 80). Possible methods include carrying out the operations on platforms on the sill beams of container cranes and the use of protected workstations on the quay.

15. Twistlock bins should not obstruct traffic routes on the quayside. However, they may be used to protect workers from traffic while twistlocks are inserted on the quay.

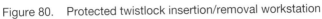

Figure 80. Protected twistlock insertion/removal workstation

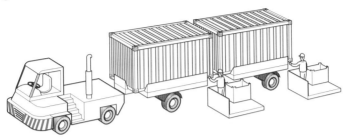

6.3.4. Changing spreader frames

1. When a spreader frame is changed:

— the work should be carried out by trained personnel;

— all electrical power circuits to the frame should be isolated before the plug is removed;

— the hoist wires on the crane should be fully slackened off before the frame is disconnected;

— the electrical plug should be stored in the dummy stowage after removal, and should not be allowed to become damp;

— frames should be securely stowed on trailers to enable them to be removed from operational areas;

— when a frame is attached, it is essential to ensure that the control switches in the cab correspond to the position on the frame.

2. If it is necessary to change a spreader frame on a crane or other container-handling appliance in a container-handling area, the area around the operation should be isolated.

3. Suitable arrangements should be made for storage of spreader frames that are not in use in a safe place that does not obstruct traffic routes. This may be on ready-use trailers.

4. Where necessary, spreader frames that are not in use should be protected by barriers and warning signs.

5. Painting spreader frames in bright colours helps to ensure that they are highly visible when kept on a quay.

6.3.5. Access to tops of containers

1. Safe means of access, such as steps, a portable ladder, a mobile elevating work platform or an access cage, should be provided if access to the top of a container is necessary. Workers should never climb up the door fittings of a container.

2. Portable ladders should not be used for access to containers stacked more than two high unless no safer means of access is reasonably practicable.

3. The surrounding area should be isolated if access is necessary to the top of a container in a container-stacking area.

4. Portworkers who have to work on top of containers should be prevented from falling off them. Whenever possible, the work should be carried out from a mobile elevating work platform or an access cage. If this equipment is not available, fall arrest equipment should be worn.

6.3.6. Operations inside containers

6.3.6.1. Opening containers

1. Sealed containers should not be opened without appropriate customs or other appropriate authority attending.

2. The doors of containers should only be opened under control (figure 81). A simple way is to first restrain them by a short sling with a spring-gate karabiner. If the door is under pressure, it will then only be able to open a short distance. It can then be opened under control by a lift truck or other restraint after the sling has been removed. If the doors are not under pressure, the sling can be removed immediately.

3. The doors of a container should be secured in the fully open position once they have been opened. This ensures the maximum natural ventilation of the container and prevents a door from being moved by wind.

4. No person should enter a container until it has been confirmed that it is safe. Hazards in addition to those of the cargo include:

— toxic gases or vapours, including decomposition products, evolved and emitted by the cargo;

— fumigant gases or fumigant residues that are still active;

— lack of oxygen.

5. No reliance should be placed on the absence of dangerous goods placards or fumigation warning signs.

6. If there is any reason to suspect the presence of a hazardous atmosphere in a container, no entry should be made until the container has been effectively ventilated and the atmosphere confirmed to be safe after allowing for any difficulty of ventilating the far end of the container. It has been found that up to 4 per cent of all loaded containers may contain dangerous levels of fumigant gases. It is therefore recommended that no container should be entered unless the atmosphere inside is confirmed to be safe.

Figure 81.   Sling to restrain container doors

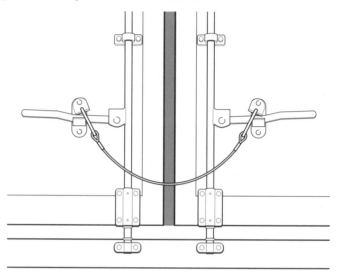

7. Sealed containers which have been opened with appropriate customs or other authority, should be resealed with an approved seal of equivalent or higher security than the original seal.

6.3.6.2. Customs inspections

1. Containers should not normally be opened for customs examinations in stacks in container parks. If it is necessary to open a container in a stack, the area should be isolated.

2. Containers to be opened for customs examination should be taken to a separate secure area with safe means of access into the container.

6.3.6.3. Packing and stripping of containers and other cargo transport units

1. Every container should be inspected before it is packed to ensure that:

— it has a valid CSC safety approval plate;

— the maximum gross weight markings on the container are consistent with those on the safety approval plate;

— it is in good structural condition, free from obvious defects and has securely closing doors;

— it is suitable for the load;

— it is clean, dry and without any residues from previous cargoes or fumigation;

— no irrelevant hazard warning placards, marks or signs have been left on the container.

2. Cargo inside a container should be packed and secured in accordance with the IMO/ILO/UN ECE *Guidelines for Packing of Cargo Transport Units (CTUs)*.

3. The load in a container should be uniformly distributed so far as this is practicable. No more than 60 per cent of the weight of the cargo should be in one-half of the length of the container.

4. A container, when packed and secured, should be sealed in accordance with relevant customs requirements.

5. Packaged dangerous goods should be segregated, packed and labelled, placarded, signed and marked in accordance with the IMO's *IMDG Code* (see Chapter 8).

6. Lift trucks used for packing or stripping a container or other cargo transport unit should be suitable for the purpose, with a short mast and low overhead guard for the operator.

To prevent build-up of dangerous exhaust gases, only lique-fied petroleum gas (LPG)-fuelled or electric trucks should be used. Lift trucks should not impose excessive point loads on the floors of containers. Container floors are generally de-signed to withstand the wheel pressure of a lift truck with a lifting capacity of 2.5 tonnes. Lift trucks with small metal wheels on the outer end of forks should not be used, as they can subject floors to high-point loadings.

7. If containers or other cargo transport units are packed or stripped while on a trailer, care should be taken to ensure that the trailer cannot move or tip up during the op-eration. Brakes should be securely applied, wheels should be chocked and the front end of the unit adequately sup-ported. Where necessary, a ramp or bridge piece should be provided (figure 82).

Figure 82.   Ramp to road vehicle

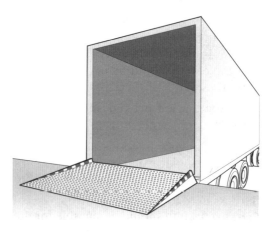

6.3.6.4. Cleaning of containers

1. All containers should be cleaned after use. In addition to basic cleaning to remove cargo residues, more thorough cleaning may be necessary for:

— quarantine control to prevent the export of pests and diseases;

— the maintenance of food quality in containers;

— the avoidance of cross-taint between successive loads.

2. All portworkers cleaning containers should be specifically trained in the potential hazards of the operations.

3. Before any container-cleaning operation is carried out, the identity of any cargo residues that are likely to be present in the container should be positively identified.

4. It should be appreciated that under some circumstances the hazards inside the container may include a lack of oxygen, in addition to the other hazards from cargo residues. Both doors of the container should be opened to maximize natural ventilation during cleaning operations, and any other necessary precautions should be taken before entry.

5. It should not be assumed that there are no hazardous residues in containers not carrying hazard warning placards.

6. Any residues found in a container should be treated as hazardous until they have been proved to be non-hazardous.

7. Appropriate personal protective equipment should be worn by all persons carrying out container-cleaning operations. The type of equipment required will be determined by the nature of the possible residues in the container.

8. If the container to be cleaned is on a trailer or chassis, safe access should be provided to and from the container. This may be by steps, ladders or other appropriate means.

9. Hazardous residues should not be swept out of containers. If they are, the relevant area will then require cleaning.

6.3.7. In-transit repairs to containers

1. Any container that is found to require detailed examination or repair should be removed from the container-handling area to a suitable area where it can be safely examined.

2. A container found on examination to have a defect that could place a person in danger should be stopped. However, if the container can be safely moved to its destination or some other place where it can be repaired, this may be permitted, subject to any necessary conditions to ensure safety, and on condition that it is repaired as soon as is practicable.

3. Damaged or defective containers should not be reloaded until any necessary repairs have been carried out.

4. Damaged or defective containers that are not to be repaired immediately should be clearly marked to ensure that they are not used.

5. Provided that it is safe to do so, a container or tank that is found to be leaking should be moved to an area where the leakage can be contained to prevent it entering a water course and until remedial action can be taken. This may involve transferring the contents to another container or tank. Some ports keep a special frame that is essentially an open tray into which a leaking container can be put and which can hold 110 per cent of the contents of the container.

6.4. Conveyors

1. A clear space of at least 1 m should be provided at each working position at a conveyor.

2. Conveyor operators should check that no persons are working on a conveyor or immediately adjacent to it before it is started. A suitable warning signal of starting should be given if the operator does not have a clear view of the whole conveyor or conveyor system. The signal may be audible, visual or both.

3. Portworkers should be prohibited from riding on all types of conveyors and from using chutes or dray ladders as means of access.

4. Rolling goods should only be moved on gravity rollers, chutes or dray ladders under the control of two ropes or other safety appliances.

5. Reception hoppers and discharge points of conveyors and transfer points between conveyors used for the movement of dusty material should be enclosed as far as is practicable.

6. The emission of dust should be prevented as far as is practicable. It can be reduced by minimizing the distance of free fall of material, by discharging through a properly adjusted spout or stocking and by the application of local exhaust ventilation.

6.5. Electrical equipment

1. Only duly authorized competent persons should be permitted to install, adjust, examine, repair or remove electrical equipment or circuits.

2. When work needs to be carried out near uninsulated electrical conductors, such as crane trolley wires, the relevant circuits should be isolated and locked off. A permit to work on a system will often be necessary to ensure that the system cannot be accidentally re-energized while the work is in progress.

3. All portable electrical equipment should be examined and tested periodically by a competent person.

4. All portable electrical equipment should be inspected at least daily by a competent person. Operators may be competent for this purpose if they have been appropriately trained.

5. Portable electric lights should only be used when adequate permanent fixed lighting cannot be provided.

6. Portable or flexible electrical conductors should be kept clear of loads, running gear and moving plant or equipment.

7. Only appropriately constructed and explosion-protected electrical equipment should be used in areas where flammable atmospheres are likely to be present.

6.6. Forest products

6.6.1. General requirements

1. All forest product cargoes should be protected from extremes of weather, as this can cause their condition, and handling and stability characteristics to deteriorate.

2. Packaged timber absorbs moisture. Weights printed on the side of packs of timber may indicate only the maximum weight of the dry timber and should only be regarded as an indication.

6.6.2. Storage

1. Storage areas for forest products should be clean, dry and level. The direction of the prevailing wind should be considered when setting the line of open-air timber stacks.

2. Stacks should be stable, uniform and well spaced to enable lift trucks to handle the widest packs safely. Isolated "tower" stacks should be avoided wherever possible (figure 83).

3. Timber should be stacked carefully on suitable bearers. Bearers should be thick enough to enable insertion of truck blades without catching the packs. The bearers of all tiers should be in a vertical line and not protrude beyond the stack (figure 83).

4. Packs that are rounded or have insufficient banding should not be stacked.

5. Stacks should be made up of timber with similar lengths whenever possible. Particular care should be taken when stacking packs of mixed length with protruding boards. Portworkers should never climb up protruding boards.

6. Stack heights should be limited to three times the width of the packs for outside storage and four times the width of the packs for inside storage. Larger block or bulk stows may be possible after careful assessment.

7. Large packs should always be below smaller packs when packs of different sizes are stacked.

8. Each tier of stacks of relatively short timber should be stacked at right angles to the tier below. The height of such stacks should be restricted.

9. Racks in which timber is stored should be marked with their maximum loads and regularly inspected for damage.

Figure 83.   Stacking of timber

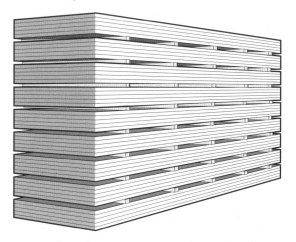

10. As cargo is de-stacked, towers should be stepped down.

11. Cargo stacks should be regularly inspected, particularly if they are old. Cargo shrinkage as moisture dries out may loosen bands and make packs and stacks unstable. Air-dried packages of timber with boards separated by small battens are particularly prone to severe shrinkage and should be inspected more frequently. Affected packs should be re-banded and re-stowed.

12. Portworkers should never climb up the sides of stacks.

13. Side-loader fork trucks may be used to obviate the need to move large amounts of cargo when picking individual packs for customer requirements. Care should be taken

when such packs are handled not to knock over the adjacent row. Pedestrians should not traverse side-loader bays when forklifts are operating.

14. The bottom tier of stacks of logs or poles should be wedged or retained by special frames to prevent slippage. The slope of the stack should not exceed an angle of 30°, smaller diameter logs or poles being placed on top.

15. Reels of paper stacked vertically on end by clamp trucks should be of the same diameter and in line vertically to ensure stability. The height of the stack should generally be not more than eight times the diameter of the reels.

16. Paper reels stowed on their side should be wedged to prevent slippage.

17. Bales of pulp can be stacked five or six bales high or greater. Each tier should be interlocked for stability. Greater stability can be achieved by "locking" second or third tiers with boards. Dunnage should be placed on the corner bales of the first tier to promote a pyramid effect of the stow.

18. Pulp should never be exposed to moisture, as it may swell, burst the banding and cause the stack to become unstable.

19. Bales of waste paper are particularly susceptible to moisture absorption and may spontaneously combust when drying out.

6.6.3. Handling

1. When lifting equipment to handle forest products is selected, allowance should be made for timber that has been

stowed unprotected or left outside for long periods, as this can increase the nominal weight of the pack significantly.

2. Particular care should be taken when handling forest products such as tongue and groove boarding or paper products, which are especially susceptible to damage.

3. Packs of timber should not be lifted above the load guard of a lift truck.

4. Lift trucks with packs elevated should only be driven short distances and at low speed.

5. Heavy braking of lift trucks should be avoided, as it may displace the load.

6. Timber stacks should be stepped as they are broken down.

7. Loose gear and dunnage should not be lifted on top of sets or packs of timber.

8. Packs should be landed on suitable dunnage bearers.

9. Loose banding wire should be collected up as work progresses.

10. When unpacked bundles of timber are lifted by a single sling, a wrapped choke hitch should be used to prevent pieces from falling out (figure 84).

11. Paper reels are normally handled by simple mechanical scissor-head clamps, hydraulic or vacuum clamps or frames, core probes or Jensen slings. The manufacturer's instructions should be followed.

12. Bundles of pulp should be lifted by large frames with quick-release hooks or flat pulp hooks, which spread the load on the wire seizings or strappings. Normal hooks

Figure 84. Single sling lift by wrapped choke hitch

may break the wires. A stretcher bar should be used to prevent hooks from sliding together under load and causing the bundle to become unstable.

13. Normally, when one lifting wire is broken, the safe working load of the utilized cargo is not met. The banding certificate should be consulted. If bundles are broken and individual bales have to be handled, suitable hooks should be used, hooks being inserted under different packing wires.

14. Pulp is normally handled by "squeeze" or bale clamps fitted to the carriage of a lift truck.

15. Bales of waste paper are handled and stowed in a similar manner to pulp.

16. Eye protection and suitable gloves should be worn when removing or replacing banding on bales of pulp or waste paper.

6.7. Gatehouses and reception buildings

1. Work near gatehouses or reception buildings should be organized so as to minimize the exposure of workers to vehicle exhaust fumes while undertaking traffic control, examination of vehicles and security duties (see sections 3.12.1 and 9.1.7).

2. The time during which workers are exposed to vehicle exhaust fumes should be limited if it is not practicable to reduce such fumes to an acceptable level. This may be done by rotating duties during the working period.

6.8. General cargo operations

1. General cargo operations should be planned so as to minimize the necessity for portworkers and vehicles to work in the same area.

2. Where practicable, walkways that have to pass through cargo-handling areas should be located at the edges of the areas, rather than passing through the middle of them.

3. All portworkers carrying out general cargo-handling operations should be provided with and should wear high-visibility overalls or other outer garments, safety footwear, safety helmets and gloves, when necessary. They should also wear any other items of personal protective equipment that may be necessary for carrying out particular operations.

4. When objects are lifted with jacks, the jacks should be:

—   constructed so that they will remain supported in any position and cannot be lowered accidentally;

—   set on solid footings;

—   centred properly for the lifts;

—   placed so that they can be operated without obstruction.

    5. If cargo platforms are used, they should be:

—   constructed of robust material;

—   sufficiently large to receive the cargo and ensure the safety of persons working on them;

—   not overloaded.

    6. Hatch covers should not be used in the construction of cargo platforms.

    7. Where heavy objects, such as loaded drums or tanks, are handled up or down an incline, their movement should be controlled by ropes or other tackle as well as chocks or wedges. Portworkers should not stand on the downhill side of the load.

    8. Drums, casks and similar cylindrical cargo that can be rolled should be kept under control at all times. They should be pushed with the hands flat on the circumference of the drum and well in from the ends in order to prevent possible trapping. Wooden casks or barrels should be pushed on their hoops.

    9. The method of stacking or storage of cargo should be determined in the light of the:

—   cargo-handling equipment that is available;

—   location and size of space that are available;

—   length of time that cargo will be kept in that location;

—   next operation.

10. Dunnage should be used as appropriate under goods that are to be loaded or unloaded by lift trucks or other lifting devices (figure 85).

11. Dunnage should be of sufficient size to allow for forks, other lifting devices or slings to be inserted or removed easily.

12. Stacks of goods should be broken down systematically from the top tier in order to ensure that the stability of the stack is maintained.

13. Where appropriate, cargo should be kept on pallets.

14. Long thin cargo, other than timber, should be kept in racks.

15. Due consideration should be given to the need to maintain the stability of racking when goods are loaded and unloaded from it. Goods should never be balanced on the edge of racking. This may lead to the overturning of racking, particularly if the lower levels of racking are empty or lightly

Figure 85.   Use of dunnage

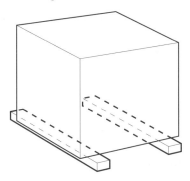

loaded and the centre of gravity of the loaded rack is above that level.

16. Cargo placed at a quay edge should be positioned so that there is a clear space of not less than 1.5 m between any part of the stack and the quay edge. If this is not practicable, the cargo should be placed in such a way that there is not enough room for a person to squeeze between the stack and the quay edge.

6.9. Machinery (general)

1. All machinery should be regularly maintained and cleaned, to ensure that it remains in a safe and efficient condition. Particular attention should be paid to the risk of corrosion that may result from the marine environment and the materials handled.

2. Any machinery that is found to be unserviceable should be isolated or immobilized, as appropriate. It should also be clearly signed or marked to show that it is unserviceable and to ensure that it is not returned to use before it has been repaired.

3. Guards of dangerous parts of machinery should not be removed while the machinery is in motion.

4. Dangerous parts of machinery should only be cleaned, examined, lubricated, adjusted or repaired when the machine is stopped.

5. Only duly authorized persons should be permitted to remove guarding from dangerous machinery. Any guarding that is removed by an authorized person should be replaced as soon as possible and before the machinery is restarted.

6. All machinery that is stopped for servicing or repairs should be isolated. Effective measures should be taken to ensure that it could not be accidentally restarted. This may be achieved by the use of lock-off devices or "permit to work" systems.

7. If it is found to be absolutely necessary to examine or adjust machinery in motion and with a guard removed, the work should only be carried out by specially authorized and trained personnel, in accordance with a safe system of work. Such persons should wear one-piece overalls with no loose ends.

8. An audible or other suitable warning should be given to workers nearby before large or complex machinery is started, unless the operator can clearly see all parts of the machine or system.

9. Pipes containing steam or other hot fluids at temperatures of more than 50°C should be suitably insulated or otherwise protected. Where necessary, pipes should be colour coded in accordance with national standards.

6.10. Mobile equipment (general)

6.10.1. General requirements

All safety-critical items of mobile equipment should be maintained in a safe and efficient condition. This should include an appropriate daily inspection by the driver, operator or other competent person, including checking tyre pressures in order to maintain stability.

6.10.2. Internal movement vehicles

1. Internal movement vehicles should only be driven by portworkers who are competent and authorized to do so.

To be authorized, they should be over 18 years of age, medically fit and appropriately trained, both on the type of vehicle used and the operations to be performed. Many port and terminal operators operate a licence or permit system that clearly identifies driver and operator competencies.

2. Seat belts or other appropriate restraints should be provided and worn where necessary. These should preferably be of the inertia reel type and designed to take into account the working position of the driver or operator, which may differ from that of a normal road vehicle.

3. Passengers should only be permitted to be carried in vehicles that are constructed to carry passengers.

4. All port vehicles should be driven within port speed limits at a speed that is appropriate for the vehicle and its load.

5. The braking capacity of a tractor unit should always be sufficient to control and stop its trailer safely when carrying the maximum payload on a loading ramp, whether or not the trailer is braked.

6. The braking system of a tractor unit should always be compatible with that of any trailer that it is to tow. Particular attention should be paid to this matter in ports where trailers from different countries with different braking systems are liable to be handled.

7. The driver of a tractor unit towing one or more trailers should:

— drive at an appropriate speed;

— not cut corners;

— allow plenty of clearance when passing other vehicles, stationary objects or people, particularly when towing a wide load;

— avoid reversing whenever possible and seek assistance when necessary;

— not reverse with more than one trailer;

— drive slowly down gradients when trailers are loaded, particularly if they are not fitted with overrun brakes;

— avoid braking fiercely, since this may cause trailers to jackknife.

8. Dusty material conveyed in port areas in open trucks should be covered to prevent it from being blown from the vehicle by wind.

6.10.3. Trailer operations

1. The braking system on a trailer should be compatible with that on the tractor unit that is to move it.

2. All brake lines and brake reservoirs should be fully charged before a trailer is moved.

3. The correct weight distribution of the load relating to each type of tractor unit used should be determined for each design of trailer used. Incorrect loading and speed are the two major causes of skeletal trailers rolling over in ports. In general:

— single 6 m (20 ft) containers should be at the rear position on a trailer. However, where there is a central stowage position on the trailer, it should be used (figure 86A). Where two 6 m containers are loaded, the heavier container (+/–2 tonnes) should be at the rear (figure 86B);

— any 12 m (40 ft) containers should be placed as near to the front of the trailer as possible (figure 86C), but 9 m (30 ft) containers should be at the rear (figure 86D). Ideally, when single 6 or 9 m containers are placed on a trailer, stop pins or ridges should prevent container movement.

Figure 86 (A, B, C and D). Distribution of load on a trailer

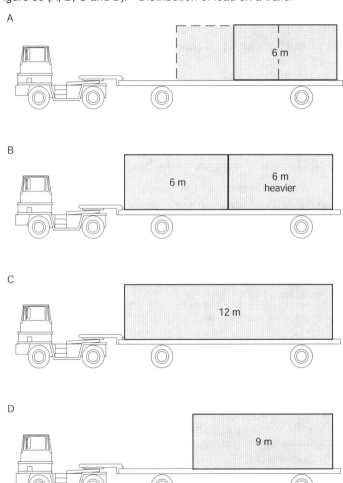

4. Trailers carrying containers that are to be loaded or unloaded from a loading bay by lift trucks should be adequately supported, by trestles or otherwise, to prevent the trailer from tipping when the truck is inside the container.

6.10.4. Trestles

1. Trestles should be moved with care, as they can cause injury if not handled correctly.

2. Trestles moved lengthways should be pushed at their mid-height, where possible, and care should be taken not to pitch into gradients or ground the base of the trestle.

3. When necessary, assistance should be sought if a trestle has to be pushed up a gradient.

4. No attempt should be made to right a trestle that has fallen over without help.

5. Trestles should be stored away from traffic routes on level ground. If it is necessary to leave a trestle on a slope, it should be chocked or tethered.

6. Special care should be taken when a trestle is placed beneath the last vehicle in a row, and when the trestle cannot be located from either side (a tunnel stow). A good system of work and signals is required, since the operation entails workers manoeuvring a trestle alongside or under a moving vehicle. Alternatively, modified trestles that can clip onto the trailer chassis, or are held or pushed by skids or bars, on the rear of the tractor unit can be used.

6.10.5. Goosenecks

1. When it is necessary to travel with an unladen gooseneck, it should be:

— preferably to the rear;

— kept approximately 1 m from the ground, to avoid damage;

— in line with the haulage vehicle and not slewed across the carriage.

2. Great care should be taken when cornering. Sharp turns can cause a gooseneck to swing violently.

3. When tractor units with goosenecks are parked, the gooseneck should be lowered to the ground in line with the vehicle.

4. Storage frames for goosenecks should be located in such a way that it is not necessary for tractor units to reverse directly into traffic routes.

6.10.6. Roll trailers

1. The loading of containers on roll trailers differs from that on skeletal trailers, in that:

— a single 6 m (20 ft) or 9 m (30 ft) container should be loaded at the front, nearest to the gooseneck end (figure 87A);

— if two 6 m containers are loaded, the heavier should be at the front (figure 87B).

2. Whenever practicable, roll trailers that are being moved on or over ramps and inclines (figure 88) should be:

— pushed (reversed) up the slope;

— pulled (forward) down the slope.

3. If containers are double stacked on a roll trailer:

— the containers should be connected by twistlocks, cones or other inter-box connectors;

Figure 87 (A and B). Distribution of load on a roll trailer

A

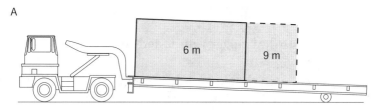

B

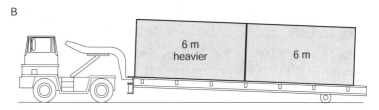

Figure 88. Roll trailer on a ramp

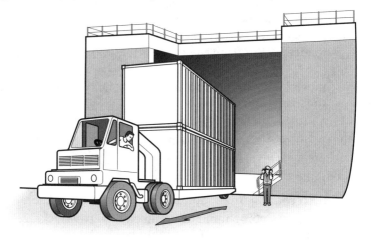

— two 6 m (20 ft) containers should never be stacked on top of a 12 m (40 ft) container;

— four-wheel drive tractor units should be used;

— travel routes should be as direct as is practicable;

— turns should be kept to a minimum. Where turns are necessary, the turning circle should be as large as is practicable;

— speed should be limited to approximately 8 kph (brisk walking pace);

— the total load should not exceed the safe working load of the trailer.

4. Double-stacked containers should only be carried by roll trailers if:

— the heavier of two 12 m (40 ft) containers is on the bottom;

— the heaviest of three 6 m (20 ft) containers is at the front of the two bottom containers and the lightest container is on the top at the rear;

— with four 6 m (20 ft) containers, the weight of either container on top does not exceed the weight of the lighter container on the bottom.

All weights referred to are gross.

5. Arrestor chains should only be necessary when towing empty or lightly loaded roll trailers with a gross weight no greater than the combined safe working load of the chains.

6. Arrestor chains should never be used when towing heavily loaded roll trailers. Should the roll trailer become detached, particularly on gradients, chains may fail and strike the tractor unit operator (see section 6.17, paragraph 2).

7. Roll trailer parks and traffic routes should be laid out to permit parking trailers with forward-facing goosenecks.

6.10.7. Cassettes

1. Cassettes should generally be operated in a similar manner to double-stack roll trailers.

2. Cassettes should always be kept as nearly in line as possible when approaching ramps or bridges. Speed should be appropriate to the prevailing conditions. Factors to be considered include the gradient of ramps, the nature of the surface and weather conditions.

3. Cargo loaded onto cassettes should be kept within the dimensions of the cassette. If it is necessary to load over-width cargoes, care should be taken not to obscure the driver's view.

4. Cargo on cassettes should be secured with appropriate strapping, angles and ratchet tensioners.

5. The lenses or protective shields of cassettes fitted with cameras at their rear should be kept clean.

6. Where cameras are fitted, cassettes may be man-oeuvred without the use of a traffic controller or banksman.

7. Cassettes are designed to take advantage of the maximum height of deck heads on ro-ro ships. Factors affecting the height and stability of the load that should be considered include:

— ratio of load height and width;
— load uniformity for lashing purposes;
— maximum headroom on board ship and elsewhere;
— gradient of bridges, ramps and 'tween ramps;

— centre of gravity of the load;

— driving speeds.

8. When cassettes are lifted, they should be kept horizontal either by lifting both ends at the same time or by lifting the ends a little alternately.

9. The tractor unit and cassette should always be stationary during loading.

10. Cassettes should never be grounded while moving. This may cause the load to shift and break lashings.

6.10.8. Parking

1. Mobile plant and equipment should only be parked on firm level ground.

2. The handbrake or other parking brake should be properly applied on all mobile plant and equipment when they are parked. Additional appropriate precautions, such as the use of chocks, should be taken to prevent unintended movement if it is found to be necessary to park on a slope.

3. Parking areas for mobile plant and equipment, including trucks and mobile cranes, should be marked out in suitable locations.

4. Mobile equipment should not be parked in a place where it is liable to cause an obstruction to traffic, or restrict the field of vision of drivers of other vehicles on roads or elsewhere.

6.10.9. Refuelling

1. Wherever possible, motor vehicles should be refuelled at fixed installations. If this is not practicable, the refuelling should be carried out in a well-ventilated space, preferably in the open air.

2. When refuelling is carried out, it is essential to ensure that:

— the engine is stopped and handbrake applied;
— the operator is off the truck;
— sources of ignition are excluded from the area;
— hot engine surfaces are protected from spillage;
— spillage and overfilling are prevented;
— any spillage that does occur is cleared up before the engine is restarted;
— filler caps are securely replaced.

3. Refuelling should be carried out in a well-ventilated place, preferably in the open air. Vehicles should not be refuelled in holds or other confined spaces.

4. LPG containers should only be changed in the open air by trained workers. The replacement container should be free of damage and all fittings should be in good order. The container should be fitted with its pressure relief valve at the top. Care should be taken to ensure that threads are not damaged and connections are gas-tight.

5. All LPG containers not in use should be kept with their relief valves uppermost in a secure, well-ventilated store.

6.11. Liquid bulk cargoes

1. Operations involving the handling of liquid bulk cargoes should be carried out in accordance with international and national industry standards and codes of practice. These include the IAPH/ICS/OCIMF *International Safety Guide for Oil Tankers and Terminals (ISGOTT)* and the

ICS/OCIMF *Safety guide for terminals handling ships carrying liquefied gases in bulk*.

2. The limits on the size of ships that may be handled at each berth should be clearly established.

3. If severe weather conditions can be experienced at the terminal, the maximum wind speed at which operations can be carried out safely should be determined. Arrangements should be made for the terminal to receive appropriate warning when wind speeds in excess of the limit can be expected.

4. Before cargo-handling operations start, the master of a tanker and the berth operator should:

— agree in writing the handling procedures, including the maximum cargo-handling rate;

— complete and sign, as appropriate, the ship/shore safety checklist;

— agree in writing the action to be taken should an emergency arise during the operations;

— agree clear unambiguous communications between ship and shore.

5. The ship/shore safety checklist should show the main safety precautions that should be taken before and during the cargo-handling operations. All items on the checklist that lie within the responsibility of the tanker should be personally checked by the tanker's representative, and all items that are the responsibility of the terminal should be personally checked by the terminal's representative. In carrying out their full responsibilities, both representatives should assure themselves that the standards of safety on both sides of the operation are fully acceptable by questioning the

other, by reviewing records and, where it is felt to be appropriate, by joint visual inspections.

6. Appropriate means of escape in case of fire should be provided from all points on the berth. This should preferably be done by the provision of alternative means of escape. If only one means of escape is available on the berth, the alternative may be by additional walkways, boat, the provision of water sprays to protect the means of escape or suitable shelters for people awaiting rescue to protect them against radiant heat or exposure to toxic gas.

7. The use of self-sealing emergency release couplings with metal cargo arms minimizes the spillage of hazardous liquids in the event of an emergency. Emergency release couplings should be used for all operations with liquefied gases.

8. Cargo-transfer hoses should be examined on each occasion before use.

9. Cargo-transfer hoses should be handled with care. Hoses should not be dragged along the ground or bent to a radius less than that recommended by the manufacturer. Where necessary, lifting bridles or saddles should be used. Hose strings should not be allowed to impose undue strain on the ship's manifolds.

10. An insulating flange or section of cargo-transfer hose should be inserted in hose strings to be used for the transfer of flammable liquids, to prevent arcing during connection or disconnection of hoses (figure 89). Care should be taken to ensure that the insulating flange or section of hose is not short-circuited. This may be by the use of uninsulated lifting cradles or otherwise. The use of a ship/shore

Figure 89.   Position of insulating flange in cargo-transfer hose

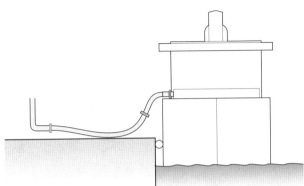

bonding cable has been found to be ineffective and potentially dangerous, and is not recommended.

11. All loading arms and pipeline manifold connections should be clearly identified to prevent accidental misconnection and resultant contamination of cargo.

12. Drip trays should be provided beneath the flanges at the end of shore pipelines (figure 90).

13. All cargo-loading arms and hoses should be drained down before couplings are opened.

14. All portworkers connecting or disconnecting cargo arms or hoses should wear personal protective equipment that is appropriate for the cargo being handled.

15. Cargo-handling operations should be kept under surveillance throughout the operation.

16. Cargo-handling operations should be stopped in the event of range or drift alarms operating and appropriate remedial action taken.

Figure 90.   Drip tray below shore pipeline connecting flange

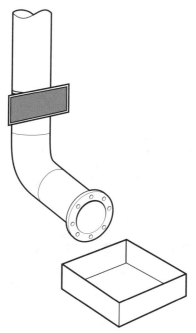

17. Cargo-handling operations should be stopped and cargo arms drained and disconnected when wind speeds approach the permitted limit.

18. No naked lights, other sources of ignition or hot work operations should be permitted at berths where operations involving flammable liquids or gases are carried out unless specifically authorized under a permit to work that en-

sures that all necessary precautions are taken in connection with the work.

6.12. Logs

1. When logs are handled, it should be assumed that they are in a saturated condition. Loose gear used to handle logs should accordingly have a safe working load that is greater than the log's dry weight by a generous margin.

2. When tongs or scissor clamps are used:

— logs should be at least 1 m shorter than the length of the hatch;

— tongs should be placed as near as is practicable to a point immediately above the log's centre of gravity; trial lifts should be carried out, where necessary, for this purpose;

— tongs should bite into the wood beneath the bark; bark should be removed at the lifting points if there is any doubt as to whether or not it has been pierced;

— persons applying tongs should stand well clear when the lift or trial lift is made.

3. Big logs should be handled with tongs slung from a lifting beam (figure 91). Tongs with claws should grip the log in the lower half of the circumference. Compression will increase with the load.

4. Wooden billets can also be slung with wire ropes. At least two slings should be used to avoid any risk of slipping. To prevent choked slings from slipping along logs, wrapped choke hitches should be used. Chains should be avoided, as it is difficult to prevent them from slipping.

Figure 91.   Log tongs

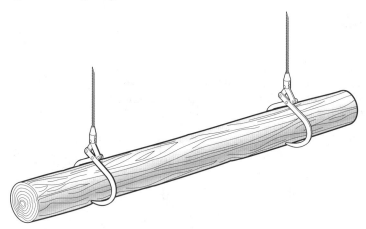

5. Special care should be taken during discharge operations in the hold if logs are stacked one above the other, as removal of one may bring others down onto portworkers.

6. Where necessary, logs should be moved under a lifting appliance by a lift truck or bull-roping so that they may be lifted vertically.

6.13. Mooring operations

1. The area around all mooring bollards should be kept clear of obstructions.

2. All portworkers carrying out mooring operations should be specifically trained. The training should preferably be coordinated with that of ships' crews. This is likely to be practicable in ports with frequent calls by regular services, such as ferry terminals.

3. The training should include the properties and hazards associated with the different types of mooring lines and the hazards of working in the "snap back" danger zone. The "snap back" zone is the area in which a person is liable to be struck by a line that parts under tension.

4. Portworkers carrying out mooring duties should wear appropriate personal protective equipment. This should normally include safety helmets, safety footwear, high-visibility overalls, buoyancy aids and gloves. If foul weather clothing is necessary, it should also be of high visibility.

5. Mooring operations should only be carried out in response to clear instructions or signals. These are made easier by the use of portable radios to communicate directly with the mooring party on the ship.

6. Members of mooring parties should stand clear of lines being thrown, but should be ready to pick them up as soon as they have landed.

7. When a winch or capstan is used to haul in a mooring line, not more than three turns should be put on the winch. The first turn should be put at the bottom of the winch. One member of the mooring party should be stationed by the winch, with a second member backing up and coiling down the slack line.

8. Members of the mooring party should only enter the "snap back" zone around a winch when it is necessary to carry out a specific task, such as releasing or dipping a mooring line. Their presence in the zone should be limited to the minimum period that is necessary.

9. No persons should put their feet in bights or eyes of mooring lines or step over a mooring line under tension.

10. The dropping of mooring lines into the water when casting off should be avoided as far as possible. Lines in the water are liable to foul bow and stern thrusters.

11. Heaving lines and other lines should be coiled down after use.

12. A "monkey's fist" at the end of a heaving line should only be made of rope. Additional weights should never be added to it.

13. When quick-release hooks are reset after use, a check should be made to ensure that they have been properly reset.

14. All mooring lines and bollards should be suitable and in good condition. Any defects in mooring lines or other mooring equipment should be reported to the appropriate person as soon as possible.

15. Small boats should not be moored to quayside ladders. Where this is unavoidable, the boats should not obstruct access to the ladders from the water and consideration should be given to the need to provide additional ladders.

16. Exercises should be regularly carried out on retrieving persons who may have fallen into the water.

17. When automated mooring systems are used, for example with high-speed ferries, the portworkers concerned should be trained in the operation of the linkspan and mooring clamps, and emergency procedures in the event of equipment failure.

6.14. Pallet handling

1. All pallets should be inspected before being used. Any pallets found to be defective should be removed from

the area and withdrawn from service. Discarded pallets should not obstruct working areas or traffic routes.

2. Pallets should be suitable for the intended load and method of handling. Many accidents have occurred when pallets have been taken from a random selection of used pallets of unknown specification. Most pallets are designed to lift a uniformly distributed load. Unless the pallet has been specifically designed for point loading, the load should be distributed as uniformly as possible.

3. Loads should be built up on pallets in an appropriate pattern so as to be compact and stable. The overlapping of individual packages, the insertion of sheets of paper or cardboard between layers, and strapping the load to the pallet by metal or plastic tapes or shrink wrapping, are some methods of increasing stability.

4. Strapping securing loads to pallets should not be over-tensioned. Deck boards can pull away from the bearers if the straps are tensioned excessively or the load is too small. Top boards should be used with small loads.

5. If pallets are to be stowed in the hold of a ship, it may be desirable for the load on a pallet to extend a short distance beyond the edges of the pallet, as this will allow them to be stowed compactly together, with little or no need for dunnage or inflatable cushions to block the stow.

6. The height of the load on a pallet should generally not exceed the longest base dimension of the pallet.

7. Pallets should not be loaded above their rated load.

8. Dangerous goods should be stacked on a pallet with their hazard warning labels clearly visible. Shrink wrapping should not be allowed to obscure such warning labels.

9. Palletized loads that are to be transported in the same condition throughout the transport chain should be conspicuously marked with the gross combined weight of the load and pallet.

10. The forks of pallet-handling devices should extend into the pallet for a distance of at least 75 per cent of the dimension of the pallet parallel to the forks. The forks should be so spaced as to ensure that maximum support is given to the pallet when it is lifted.

11. The small finger wheels of pallet trucks should not be allowed to damage base boards of pallets.

12. Pallets handled by crane should only be lifted by suitable fork attachments or, in the case of wing pallets, by bar slings with spreaders.

13. Loaded pallets which on visual examination do not meet the requirements set out above should be secured onto pallets in good condition before being further handled. Alternatively, the load can be removed and placed on a satisfactory pallet.

14. Empty pallets should be handled with care and should not be dragged or thrown down. They should not be handled by methods liable to damage or loosen them, such as the insertion of the platform of a sack truck between the bearers.

15. Pallets should never be used with a forklift truck as an improvised access platform.

16. Pallets that are not in use should be kept in appropriate designated places that are protected from the weather.

6.15. Passenger terminals

1. Access by passengers should be limited to appropriate areas physically segregated from all other operations, whenever this is practicable. Areas to which access by passengers is prohibited should be marked.

2. Passenger movements should be monitored and controlled to ensure that they remain in permitted areas.

3. Vehicle access routes and passenger drop off/pick up areas should be monitored and supervised, to ensure an effective and efficient flow of traffic.

4. Additional signs and other traffic control measures should be provided as necessary during visits by cruise ships, when unusually large numbers of vehicles may be present for short periods. Whenever possible, passenger traffic should be segregated from other traffic, including that provisioning the ship.

6.16. Rail operations

6.16.1. General requirements

1. So far as is appropriate, port railways should be operated in accordance with the principles of the rules for similar lines in the national railway system.

2. Written procedures, which may involve specialized training for the operation of transfer yards and other rail operations that may be carried out in conjunction with workers of the national rail system or other organization, should be drawn up and agreed between the relevant organizations.

3. Written safe systems of work should be developed for port railway operations. These are likely to include:

— shunting (switching) by any means;
— coupling and uncoupling;
— braking and brake testing;
— operation of level crossings;
— liaison with other rail organizations for the dispatch and receipt of trains;
— derailment procedures;
— track maintenance.

4. No person should pass under or between railway vehicles.

5. No person should cross railway lines within 15 m of a stationary railway vehicle, unless it has been positively established that it will not move. Particular care should be taken when crossing behind a train that has just stopped. Braking may have taken up slack between vehicles and compressed couplings and buffers. Wagons may then roll back several metres after the train stops.

6. All persons working on or near railway lines should wear high-visibility clothing.

7. All persons driving locomotives should be duly authorized to do so, medically fit and appropriately trained.

8. Locomotive drivers should only act on clear signals given by an authorized person. However, an emergency stop signal to stop should always be obeyed, irrespective of the source.

9. Locomotive whistles should be sounded or other appropriate warning devices actuated before locomotives or trains move off, and on approaches to level crossings and other hazardous places.

10. Unauthorized persons should never ride on locomotives or railway wagons.

11. Locomotives passing through a port area where people are working should move at dead slow. When wagons are pushed through such areas, the front wagon should be accompanied or preceded by a person on foot who is clearly visible to the driver at all times.

12. If a radio is used to communicate with a locomotive driver to direct a shunting or other rail movement, continuous radio contact should be maintained, so that the integrity of the communication is constantly monitored and confirmed. This may be done by continuous repetition of the word "proceed".

13. Care should be taken during shunting operations to ensure that both facing and trailing points are correctly set before the start of the movement. Level crossings and unprotected openings near the track should be supervised as appropriate.

14. No worker should climb above a footplate or floor level on any locomotive or wagon while under overhead electrified lines.

15. No goods or other obstruction should be placed within 2 m of the nearest rail of a track.

16. No wagons should be parked beyond the fouling point at the end of a siding.

17. Wagons and other vehicles should not be parked close to road or pedestrian crossings or other place where they are liable to cause an obstruction or obscure the view of road or rail drivers.

18. Brakes should be applied to parked rail vehicles to prevent unintended movement. Where necessary, the wheels of a vehicle may be scotched. Scotches should only be applied to the wheels of stationary vehicles.

19. Work on railway tracks should only be carried out if appropriate steps have been taken to protect those carrying out the work from the approach of trains. This may be by a permit to work giving total possession of the section of track, by the provision of appropriate warnings and the use of lookouts, or by other appropriate means.

20. No worker should be permitted to work between or beneath railway vehicles unless positive steps have been taken to prevent the movement of the vehicle or vehicles and the approach of other vehicles.

6.16.2. Loading and unloading of rail wagons

1. When opening wagon doors, portworkers should check that the door fastenings are in good order and stay clear of the door and any goods that may fall as the door is opened.

2. Workers should not be inside open wagons (gondola cars) when bulk material is handled by grabs or magnetic lifting devices.

3. Danger signs should be placed at either end of sections of passageways or walkways near which open wagons with swinging side doors are being emptied.

4. Precautions to prevent workers' fingers from being crushed should be taken when drop doors on hopper wagons or open wagons with hopper bottoms are opened.

5. Before wagons are moved, all hinged doors should be properly fastened, and insecure and overhanging stanchions or metal straps should be removed and placed at least 2 m clear of the rails.

6. Bridge plates between loading platforms and the beds of wagons should be properly secured. When not in use, they should be kept in a suitable location at least 2 m from the edge of the platform.

7. Suitable tools should be provided and used for unfastening metal straps.

8. Lift trucks should not be used inside railway wagons unless the floors of the wagons are in a safe condition.

9. Double-deck wagons that are to be loaded or unloaded should have handrails at the sides of the top deck. Walkways adjacent to the handrails should have a non-slip surface.

6.16.3. Moving rail wagons

1. Train crews should ensure that all portworkers are out of the wagons and all persons clear of the area before moving railway wagons.

2. Railway wagons should only be moved under proper control. Fly shunting (flying switches) should be prohibited in port areas.

3. Wagons moved by a locomotive should normally be coupled to it.

4. The number of wagons moved by a locomotive should not exceed the number that can be controlled by the brake power of that locomotive.

5. Wagons should be stopped only by applying the brakes. If it is necessary to move a wagon that is not coupled to a locomotive, the wagon should be under the control of a brakeman.

6. All trains should be brought to a full stop before any wagons are uncoupled.

7. Cranes, ships' winches and derricks should not be used to move railway wagons.

8. The use of locomotives or wagons as a ram to adjust a load on a wagon should be prohibited.

9. Where it is necessary to move railway wagons more than a short distance without a locomotive, the wagons should be moved by means of power-driven wagon movers, winches or capstans.

10. Power-driven wagon movers should be driven to the side of the track when pushing wagons so that the driver has the best possible view. When pulling a wagon, the driver should be able to disconnect the traction rope from the driver's seat in case of danger.

11. When a railway wagon or a group of wagons that is not coupled to a locomotive is being moved, a worker should control each wagon or group of wagons. Unless the person in control of the wagon has a clear field of vision, the operation should be directed by a signaller in a safe position.

12. If wagons are moved by capstans:

— treadle-operated capstans should be tested each day before use;

— areas immediately around capstans should be kept free of obstructions;

— capstan controls should be on the side away from the live rope and located so that the operator is clear of the intake of the rope;

— wagons should only move under control;

— synthetic fibre rope should not be allowed to heat by friction on the capstan;

— workers should not stand between the rope and the wagons, and should keep clear of the hauling rope.

13. Ropes or hawsers used with capstans should be regularly examined.

14. Wagons should not be moved by means of push poles and locomotives or wagons on an adjacent track or crossover.

15. When it is necessary to move wagons without mechanical power, workers should never:

— push wagons when standing between the buffers of two coupled wagons;

— press on buffers with shoulders or hands to push vehicles;

— push a vehicle by putting hands on slideways of doors, frames of open doors or open doors;

— stand in front of a moving vehicle;

— slow down a vehicle by pulling on the buffers.

16. Wagons should not be left standing on quays with short distances between them unless this is necessary for operational reasons.

17. Damaged wagons should only be moved when necessary and with extreme caution, especially if a coupling or buffer is damaged or missing.

6.17. Roll-on-roll-off (ro-ro) operations

1. Any necessary checks on ro-ro vehicles and their cargo should be carried out at the access gate or other suitable place.

2. Particular attention should be paid to any couplings between vehicles to ensure that they will not become uncoupled on a ship's ramp. Particular attention should be paid to vehicles towing caravans, which should always use proper ball hitches and trailers. Goosenecks on tractors are liable to become disconnected from a trailer at the ends of a ramp if the slope is too great. Additional side safety chains or other fastenings should be used where necessary except when heavily loaded roll trailers are being stowed (see section 6.10.6, paragraph 6).

3. Checks on the declaration and placarding, marking and signing of dangerous goods should be carried out in accordance with national legal requirements.

4. Abnormal loads may need to be escorted directly to or from the ship or a suitable waiting area.

5. Ro-ro traffic should be appropriately controlled at all times. All traffic marshals should wear high-visibility clothing. Speed limits should be enforced.

6. Parking on ro-ro traffic access routes should be prohibited except in suitable designated areas. Vehicles carrying dangerous goods in such areas should be segregated as necessary.

7. Adjustments of loads on vehicles and the sheeting and unsheeting of loads on vehicles should only be permitted in clearly indicated designated areas.

8. Trailer legs should be lowered to the ground before trailers are uncoupled. It is important to ensure that the trailer parking brakes have been properly applied and the front of the trailer left high enough to permit another vehicle to couple to it.

9. The shore approaches to ramps of ro-ro ships should be kept clear at all times.

6.18. Scrap metal

1. The effects of possible noise and dust on neighbouring premises and activities should be considered when a quay is selected for scrap metal operations.

2. The maximum permissible size of piles of scrap metal at quays where it is handled should be determined.

3. A clear accessway should be left between the edge of scrap metal piles and quay edges.

4. Portworkers should be alert to the possible hazards of scrap metal received. These hazards include the following:

— flammable residues inside closed vessels;
— lack of oxygen in closed receptacles or containers due to rusting or other atmospheric oxidation;
— the presence of radioactive sources or radioactive contamination in scrap from demolition or dismantling of plant at factories and mines;
— heating of consignments of aluminium smelting by-products or turnings that have become damp.

5. Magnetic lifting gear should be used in accordance with section 5.3.4, paragraphs 20 and 21.

6. Where ships are loaded with scrap metal directly from vehicles, a substantial barrier should be provided to prevent the vehicles from accidentally going over the edge of the quay.

6.19. Solid bulk cargoes

1. Solid bulk cargo-handling operations should be carried out in accordance with the IMO's *Code of Practice for the Safe Loading and Unloading of Bulk Carriers (BLU Code)*. Compliance with the *BLU Code* is being made a legal requirement in a number of countries. Grain should be loaded in accordance with the IMO's *International Code for the Safe Carriage of Grain in Bulk (International Grain Code)*.

2. A port or terminal where a bulk carrier is to load or unload should appoint a terminal representative to have responsibility for the operations that are to be carried out by the terminal in connection with that ship.

3. Copies of relevant terminal and port information books should be given to the master of a bulk carrier, if possible before arrival. This may be done by electronic means. The recommended contents of port and terminal information books are set out in Appendix 1 of the *BLU Code*.

4. The terminal representative should ensure that appropriate information about the bulk cargo to be loaded is given to the ship. This should include specification of the cargo, its stowage factor and angle of repose, its moisture content and chemical properties if relevant, and trimming procedures. A recommended form for cargo information is set out in Appendix 5 of the *BLU Code*.

5. The terminal representative should agree the ship's loading or unloading plan with the ship's master. An example of such a plan is given in Appendix 2 of the *BLU Code*.

6. The terminal representative and the ship's master should jointly complete and agree a ship/shore safety checklist before loading or unloading is started. The checklist is set out in Appendix 3 of the *BLU Code* and guidelines on its completion are given in Appendix 4.

7. Loading and unloading operations should only be carried out in accordance with the plan. Any change that is found to be necessary should be agreed upon by both the terminal representative and the ship's master.

8. Planning of storage areas for solid bulk materials should take into account the angle of repose and other relevant properties of the material. Allowance should be made for any alteration that may be caused by events, such as vibration, impact or alteration of the moisture content, that could lead to a collapse.

9. Appropriate measures should be taken to suppress dust that could result from cargo-handling operations. The measures will depend on the properties of the material and individual factors. Measures could include water sprays, local exhaust ventilation at loading and unloading points, and covering the material and keeping it inside buildings, silos or hoppers.

10. Bulk material should not be kept against walls of buildings or elsewhere unless it has been confirmed that the walls have sufficient strength to withstand the maximum horizontal pressure to which the material may subject them.

11. Silos, hoppers and storage bins should be designed with smooth sides to allow discharge without hang-up of material on the sides. Where appropriate, vibrators should be fitted to ensure that any residual material is dislodged. Any necessary cleaning operations should be able to be carried out from outside the silo or hopper, whenever this is practicable.

12. Conveyor systems carrying material that is likely to produce dust should be enclosed to protect the material from wind.

13. The loads of open vehicles carrying solid bulk material should also be covered to prevent stripping of the cargo by wind. A sheeting system should preferably be built into the vehicle and be able to be operated from the ground.

14. A clear signal should be given to all persons in the area before conveyor systems are started.

15. Regular cleaning should be carried out to prevent the build-up or accumulation of dust.

16. The interior of all buildings containing silos of grain, animal feed products and similar flammable material should be regularly cleaned in order to prevent any secondary explosion in the event of failure of a conveyor system or other incident. Much serious damage that occurs in the event of a dust explosion is caused by secondary explosions of dust blown up by the primary explosions.

17. Portworkers should only be permitted to enter silos, hoppers or storage bins for cleaning, clearing a blockage or other purposes under controlled conditions. Many workers have been asphyxiated after sinking into free-flowing solid bulk materials. Entry should normally be con-

trolled by a "permit to work" system. The permit to work should ensure that:

— no further material will enter the silo or hopper;

— no discharge valve will open or conveyor start up;

— the atmosphere in the hopper or silo is safe to breathe;

— the worker is wearing a suitable harness connected to a lifeline or other suitable means for rescue in an emergency;

— one or more workers immediately outside the silo or hopper are aware of the action to be taken in an emergency and are capable of carrying it out.

18. When material is present in the silo or hopper, lifelines should be kept as taut as is practicable without impeding movement. Where possible, lifelines should be belayed at suitable positions so that if a worker loses his foothold or material collapses his weight will be immediately supported.

19. It should be appreciated that some solid bulk materials that can be kept safely in small quantities may become hazardous if kept in large stacks.

6.20. Stacking and stowing of goods

1. Goods and materials that are not in containers or vehicles should be kept in stable and orderly stacks or piles on ground or floors that are firm and level.

2. Matters to be considered when determining the method by which the goods are stacked or otherwise kept should include:

— maximum permissible loadings of quays or floors;

— possible presence of underground sewers and culverts;

— types of mechanical handling equipment available and the space in which it will be used;

— whether the goods are classified as dangerous goods;

— the shape and mechanical strength of the goods and their packaging;

— length of time the goods or material will be kept;

— the natural angle of repose of bulk material.

3. Stacks of goods should remain stable at all times. With the increased reach of modern cargo-handling equipment, the maximum height of a stack is likely to be determined by the need to ensure that the stack remains stable. The height of stacks should generally be limited to 6 m.

4. Loaded pallets should not normally be stacked more than four high. The use of pallets with uniform or generally similar loads makes it possible to form stable stacks of a simple shape that can easily reach a height of 4 or 5 m.

5. Stacks should not be of such a height or shape as to be unstable in high winds.

6. The area of an individual stack should be limited to 450 m^2 if the fire risk is small, or 150 m^2 if the goods in it would burn fairly easily. Passageways at least 3 m wide should be left between stacks to enable the use of appropriate cargo-handling equipment and to form firebreaks.

7. Stacks of goods should be broken down systematically from the top tier in order to ensure that the stability of the stack is maintained at all times.

8. Dunnage should be placed under goods that are to be loaded or unloaded by lift trucks or other lifting devices as necessary (figure 92). The dunnage should be of sufficient size to allow forks or slings to be inserted or removed easily.

Figure 92.   Use of wedges to restrain cargo

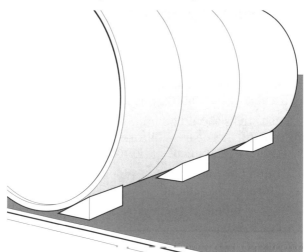

6.21. Steel and other metal products

6.21.1. General requirements

The differences in the properties of metals commonly shipped through ports should be taken into account when determining appropriate methods and equipment for their storage and handling. For example, lead is more than five times as heavy as aluminium, both are far softer than steel, and all metals have a low coefficient of friction.

6.21.2. Storage

1. Special consideration should be given to permissible floor loadings of warehouses and quaysides. If necessary,

additional bearers should be used to spread the load, particularly on lower tiers.

2. Hardwood chocks or bearers are preferable to softwood. Wooden railway sleepers are ideal for bottom bearers.

3. Tiers of long sections or plates should be separated by suitable dunnage, bearers or chocks. These should be placed in a vertical line above one another.

4. Push sticks should be used when necessary to adjust the position of the bearers. Workers should never insert their hands between tiers of stacked material.

5. Round bars or billets and pipes should be wedged to prevent lateral displacement.

6. Bars or billets of small cross section are best stored and handled in bundles.

7. Flanged pipes should be turned "end for end" in alternate tiers to protect the flanges and prevent damage. Spacers are necessary on the bottom tier to separate the flanges.

8. Wedges and spacers in stows of round bars or pipes should be of suitable dimensions and preferably nailed to bottom bearers to prevent displacement.

9. Climbing on stows should be avoided wherever possible.

10. Plates or heavy sheets stacked on edge should rest on a solid foundation and lean against vertical supports of adequate strength.

11. The bottom tier of coiled sheet steel stacked in nested tiers, i.e. with the coils in the tier above resting in the hollows formed by the coils below, should be firmly chocked.

12. Coils of varying diameter should be stacked in descending size of coil, with the largest diameter at floor level. The height of the stack should be limited to not more than five tiers.

6.21.3. Handling

1. Lifting equipment to handle steel or other metal sections or fabrications should be selected carefully, as it is often large and awkward to handle. Consideration should be given to such factors as:

— size, shape and weight of the load;

— weight of loose gear to be used;

— control of the load;

— size of handling areas and any travel routes.

2. Long steel sections should be handled by cranes or side-loading lift trucks. Forklift trucks are not suitable as the load is liable to be unstable and difficult to control, particularly if sudden braking is necessary.

3. Long sections should be slung in two places, preferably from a lifting beam, to make it easier to balance the load and prevent accidental sliding or unhooking (figure 93). Sliding of slings can also be prevented by using wrapped choke hitches or wedging slings with wooden chocks (never metal on metal).

4. Great care should be taken when lifting metal with forklift trucks. In particular, the operator should ensure that:

— truck forks are as far apart as possible to provide maximum stability;

— travel speed is reduced as necessary;

— heavy braking is avoided.

Figure 93.   Use of lifting beam with double wrapped slings

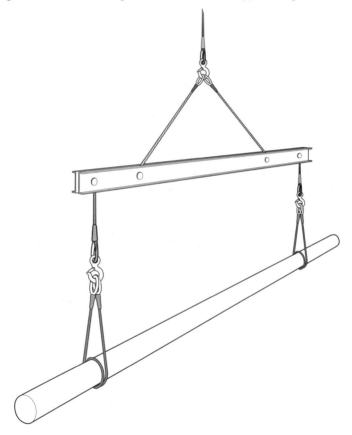

   5. If metal products are wrapped in oiled cloth or other materials to prevent corrosion, it may be necessary to provide a non-slip surface to the blades of the truck, such as rubber or dunnage flats secured to them.

6. Protection to prevent chafing and damage may be necessary between metal products and lifting equipment. Rubber strips, padding or coir mats may be suitable, depending on the load.

7. Stretchers and beams should be used where necessary. These should be at least one-third of the length of the load.

8. Long sections, particularly thin plate, are prone to bending or sagging and should be supported at more than two points. Pieces of long thin plate should be stabilized for lifting by a strong-back on top of the plate tightened by chain slings.

9. Tag lines should be used where necessary to control loads handled in gusting winds, tight stowage areas or loads that spin when lifted.

10. Special equipment, such as pipe hooks, clamps, vertical or horizontal plate clamps, coil probes, etc., should be used when appropriate. When such attachments are used, care should be taken to ensure that the load is well within the safe working load of the attachment and other loose gear, and that the combined weight of the load and loose gear does not exceed the safe working load of the lifting appliance.

11. Plates hanging vertically on edge should only be handled by self-locking plate clamps. The clamp should not be directly attached to the hook of the lifting appliance but should be connected to it by a short length of chain. Only one plate at a time should be handled in this way.

12. When two vertical plate clamps secured to a two-legged sling or two separate slings are used to raise or lower

a plate by gripping one edge of a vertically hanging plate, the clamps should be applied so that their centre lines are in line with their respective sling legs (figure 94).

Figure 94.   Use of vertical plate clamps

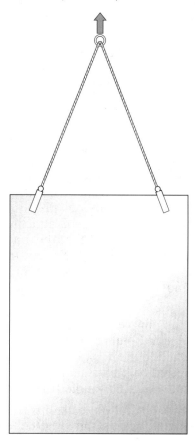

13. More than one plate may be clamped when horizontal plate clamps are used, provided that the clamps are kept at 90° to the edge of the plate, if necessary with a lifting beam.

14. Special camlock lifting clamps should be used to handle railway lines. Long sections should be supported at more than two points.

15. As road and rail transport often carry steel sections or bundles on their flat side for stability, it may be necessary to turn them under careful control to prevent dunnage breaking in stows. This may be done by turning in rotating cradles or slinging with offset slings and carefully lifting, slewing and lowering once the weight has passed its centre of gravity.

16. A trial lift, to check that slings are adjusted to balance the load, should be carried out when awkward or irregular-shaped loads are handled.

6.22. Trade vehicles

1. Trade vehicle traffic should be strictly controlled to ensure the safety of all persons in the area. Although many trade vehicles are cars (automobiles), they can also include a wide variety of other vehicles and plant that may be large, cumbersome, difficult to manoeuvre and complex. Port calls by car carriers are likely to require the movement of large numbers of vehicles over a short period.

2. Trade vehicles should normally be loaded or discharged separately from other ro-ro traffic. Car carriers are generally dedicated ships, but heavy vehicles and specialized plant are often carried on other ships. In such cases, the

trade vehicles should normally be loaded before the general ro-ro traffic and unloaded after it.

3. Holding areas for trade vehicles should be as close to the berth as is practicable, and laid out to avoid or minimize the need to reverse into parking spaces. Traffic routes within the holding area should be one-way.

4. Traffic routes between the holding area and the ship should preferably be separate from those for other traffic. Separation may be physical, by the provision of separate routes, or by timing, by restricting access by other vehicles during the loading and discharge of the trade vehicles.

5. Only authorized portworkers should drive trade vehicles. Before such workers are authorized, it should be confirmed that they are medically fit to drive the relevant type of vehicle and competent to do so. In many cases, the holding of a national licence for a given class of vehicles should be a minimum requirement. It should be noted that some licences may be limited to vehicles with automatic gearboxes, to certain vehicle sizes or in other ways. Additional training and instruction may be necessary before authorization is granted to drive certain types of vehicle. This may be available from the vehicle manufacturer or local representative. In all cases, trade vehicle drivers should be fully instructed about the type and relevant characteristics of vehicles to be driven.

6. Seat belts fitted to trade vehicles should be worn by drivers.

7. Trade vehicle drivers should observe all speed limits and keep a safe distance from the vehicle in front at all times. Trade vehicles may need to be subject to lower speed limits than the normal speed limit applied in ports.

8. Drivers of trade vehicles should be taken back to the holding area or ship after delivering a vehicle by a suitable van or other vehicle. This both separates pedestrians from vehicles and speeds up operations. The van should normally accompany the convoy of vehicles to or from the ship. The number of vehicles in the convoy, or section of the convoy, should be limited to the capacity of the van.

9. Additional precautions to protect drivers from the hazards of vehicle fumes in the holds of ships and other enclosed places may be necessary when vehicles are shipped as an incomplete chassis or are poorly maintained (see section 9.1.7).

6.23. Traffic control

1. Road and rail traffic in port areas should be controlled to ensure the safety of all persons on the premises.

2. Where appropriate, entry to all or specified parts of port areas should be restricted to authorized traffic.

3. National legal requirements relating to road and rail traffic should generally be taken as the basis of standards at ports.

4. Consideration should be given to the enactment of appropriate by-laws where national traffic laws do not apply on port premises. The by-laws should include powers of enforcement. These should be used whenever necessary.

5. Vehicles used in ports should generally comply with national standards for such vehicles. This will already be required if they are also used in the national traffic system.

6. All drivers of vehicles in ports should be authorized to do so, medically fit and appropriately trained. In general,

drivers should be required to hold the appropriate licence to drive similar vehicles in the national traffic system and to receive appropriate job-specific training to carry out specified operations.

7. Appropriate speed limits should be set on all port premises. Different limits may be appropriate in different areas. In many cases, the limits will need to be lower than national limits. The speed limits should be clearly indicated, particularly at any point where the limit changes.

8. Suitable parking areas should be provided to ensure that waiting vehicles are not a hazard to other traffic. Such hazards are most likely to arise from vehicles obstructing or restricting the vision of other drivers. Parking should be prohibited near road junctions, road or rail pedestrian crossings, sharp bends or other hazardous areas.

9. Traffic signs should conform to national standards.

10. At ferry terminals with ro-ro services to countries that drive on different sides of the road, signs should remind drivers of the correct side of the road to drive on. These signs should be multilingual where necessary.

6.24. Warehouses and transit sheds

1. Floors of warehouses and transit sheds should not be overloaded. Any restrictions on the amount or types of goods that may be kept should be clearly indicated by notices.

2. All goods in warehouses and transit sheds should be kept in an orderly fashion. Obstructions should never be left in aisles. Rubbish and dunnage should be removed and safely disposed of as soon as is practicable.

3. Stacks of goods should be separated by aisles of sufficient width to permit the safe use of lift trucks and other cargo-handling plant that is likely to be used in the building. The edges of the aisles should be clearly marked.

4. Where practicable, a one-way traffic route should be operated in the building. The traffic direction should be clearly indicated.

5. All goods should be stacked in such a way as to be stable. Where necessary, safe means of access should be provided. Only properly constructed access equipment should be used for this purpose. The forks of a lift truck or the hook of a crane should never be used. Access should never be permitted near bare conductor wires for overhead cranes.

6. When work is being carried out on a high stack, warning notices should be displayed and adequate measures taken to ensure the safety of persons passing below.

7. Suitable handholds should be provided if railings or other fencing for openings in walls or floors are likely to have to be opened or removed to permit the passage of goods. The railings or fencing should be closed or replaced as soon as is practicable.

8. Where there is danger from overhead bare crane conductor wires or from other electrical equipment, or a risk of trapping by equipment such as overhead travelling cranes, the height of stacks should be limited.

9. Any potentially dangerous conditions found in warehouses or transit sheds should be reported immediately and appropriate action taken.

10. Portworkers working in climate-controlled warehouses should be provided with suitable personal protective clothing. Those working in such areas for long periods should be permitted to take breaks in normal areas at appropriate intervals.

6.25. Confined spaces

6.25.1. General requirements

Serious risks to the health of portworkers can arise from entering or working in confined spaces. The term "confined space" means an area that is totally enclosed. It does not mean airtight, nor does it refer just to a small space. While small spaces can be confined and potentially dangerous to enter, the risks also apply to much larger spaces. A ship's hold may be a large void but, with the hatch covers on, it is a confined space and the atmosphere in it may well be hazardous.

6.25.2. Hazards and precautions

1. The normal atmosphere that we breathe is composed of 79 per cent nitrogen and 21 per cent oxygen, and it is the latter that sustains life. Variations in that amount, either up or down, can have serious effects.

2. If extra oxygen is fed into the atmosphere in a confined space, it will increase the flammability of materials and widen explosive limits, making ignition leading to serious fires or explosions more likely. If such a situation develops, the oxygen or enriched air supply should be cut off and the confined space thoroughly ventilated before any further work is undertaken.

3. If the oxygen content of air is reduced, its ability to sustain life is also reduced, and at 16 per cent or below it will

fail to sustain life at all. A portworker entering such an atmosphere without adequate respiratory protection will rapidly become unconscious and may die.

4. Lack of oxygen in a hold or other confined space may result from:

— absorption of oxygen from the atmosphere by the cargo. This can happen relatively quickly. A closed hold full of copra, for example, needs only a few hours to reduce the oxygen content to a dangerously low level. A range of bulk cargoes can behave similarly;

— rusting or other oxidation of the space or the cargo, such as scrap metal, held within it;

— decomposition or rotting of cargo;

— gas cutting or welding operations.

5. Toxic or flammable gases can also build up to dangerous levels in confined spaces. This can result from the decomposition of coal, fishmeal, bark or other cargoes, the leakage of packages due to faulty filling, packing or transit damage, vehicle exhaust fumes and leaking pipes, hoses, etc.

6. Although there are potential confined spaces in warehouses and elsewhere in port areas, accidents during port work are most likely to occur on board ship, particularly when workers enter holds.

7. Entry into a confined space should not be permitted unless:

— the space has been adequately ventilated. The degree of ventilation required will depend on the size of the space, the likelihood of the air being contaminated and the possibility of the contamination continuing (compressed air should never be used for this purpose);

337

— a competent person has declared it safe to enter.

8. If there is any doubt, the oxygen content and toxic or flammable gas concentrations should be measured before unrestricted entry is allowed.

9. If unrestricted entry cannot be allowed, a responsible person should take control and only permit entry of specifically trained and instructed workers:

— under close and adequate supervision;

— subject to a "permit to work" system;

— wearing appropriate respiratory and other personal protective equipment;

— if suitable rescue arrangements have been made.

7. Operations afloat

7.1. General provisions

1. This part of the code applies to operations that are carried out solely on board ships. Chapter 6 applies to operations that take place both on ships and on shore.

2. All portworkers, including supervisors, on board ships should be fully trained and competent. This is essential, given that more accidents involve portworkers working on board ships than at any other location in ports.

3. All port operations on board ships should be carried out in accordance with safe systems of work. These should be drawn up following identification of the hazards, assessment of the risks and development of measures to control them.

4. Experience has also shown that regular inspections and reports on the condition of ships will help to reduce the number of shipboard accidents involving portworkers. If deficiencies affecting the safety of portworkers persist on a particular ship and are not corrected after more than one voyage, a report may be made to the national competent authority inspector or port state control officer.

5. It is the responsibility of the ship to provide conditions on board in which port work can safely be carried out. However, before starting to load or unload a ship, the company responsible for the stevedoring work (the "stevedore") should itself take steps to ensure that:

— there are safe means of access onto and about the ship (see sections 7.2 and 7.3);
— a ship's lifting appliances and lifting gear (if they are to be used for cargo operations) are correctly certificated,

and appear to be in good order and safe to use (see sections 4.2 and 5.1);

— suitable deck and under-deck lighting, with a minimum level of 10 lux on access routes and 50 lux in working areas, taking into account any specific need that may require additional lighting, is provided;

— slings around pre-slung cargo on a ship have been certificated and are in all respects safe to discharge the cargo;

— any lashing gear to be used is suitable, in a safe condition and compatible with the cargo to be lashed.

6. If it is found that the provisions made by the ship are not safe or do not comply with international legal requirements, the deficiencies should be reported to the ship's master or his/her representative, the ship owners and the shore-side management. The stevedore should not allow work to start until the deficiency has been corrected. Alternatively, a shore-side provision may be made which remedies the situation, for example:

— a shore-side gangway is provided;

— quay cranes or other shore-based lifting appliances are used where possible;

— portable lighting is provided by the stevedore;

— stevedore slings are placed around the cargo.

7. A stevedore who proposes to handle cargo with ship's gear should verify that the gear is safe by checking certificates and carrying out visual inspections before it is used.

8. Any shore-side appliances and gear used on board a ship should fully comply with Chapters 4 and 5.

9. The same standards of housekeeping that apply on shore should be applied to those parts of the ship used by portworkers.

10. Any damage to the ship or its equipment that occurs during cargo handling or other activities should be immediately reported to a responsible ship's officer.

7.2. Access to ships

7.2.1. General requirements

1. Sufficient, safe and suitable means of access to the ship should be available for the use of portworkers passing to and from the ship. The means of access should be of sound material and construction and adequate strength, be securely installed and maintained in a good state of repair. Means of access should, wherever possible, be constructed in accordance with international standards.

2. The means of access from the quay to the ship's deck should be the ship's accommodation ladder, whenever this is reasonably practicable.

3. If the use of an accommodation ladder is not reasonably practicable:

— a gangway may be used;

— when normal access equipment cannot be used owing to the ship's high freeboard,[1] purpose-built shore-side access equipment should be provided and used;

---

[1] For the purposes of this code, "freeboard" means the height above water level of the deck used for access via the ladder when the access is used for the first time.

Figure 95.   One means of access: A gangway and a safety net

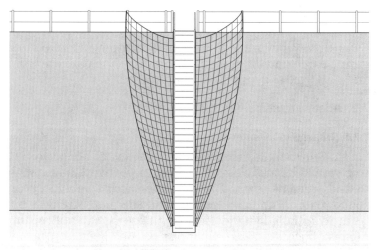

— where the freeboard is too low for the normal means of access to be used, the ship or barge should be moored alongside a quayside ladder (see section 3.3.5);

— portable ladders should only be used where no safer access is reasonably practicable;

— where the access is to/from a ship and a barge or other vessel of low freeboard moored alongside it, a rope ladder may be used when it is impracticable to comply with any of the above requirements.

4. The means of access should:

— be so placed as to ensure that no loads pass over it; if this is not practicable, it should be supervised at all times during cargo handling;

— be placed where access to it will not be obstructed;

— not be placed on or near a crane track, railway track or other route in the port where it could be struck by moving traffic on that track or route.

5. A safety net should be rigged wherever a person may fall between the ship and the quay from a means of access to a ship (figure 95). So far as is reasonably practicable, the net should protect the entire length of the means of access.

6. The relationship between the quay and the ship is not always static. When necessary, the means of access should be regularly checked to ensure that it is correctly adjusted. The master of the ship should appoint a person for this purpose.

7.2.2. Accommodation ladders

1. A ship's accommodation ladders should be set in a safe position and safety nets deployed (figure 96).

2. The construction of the ladder should be sufficiently robust to reduce any sway or bounce to a minimum. It should be fenced on both sides along its entire length with both upper and intermediate guard rails.

3. The ladder should be properly rigged and be kept adjusted in such a way that:

— whatever the state of the tide or the draught of the ship, the ladder's angle to the horizontal does not exceed approximately 40° as far as this is practicable;

Figure 96.   Accommodation ladder with a safety net

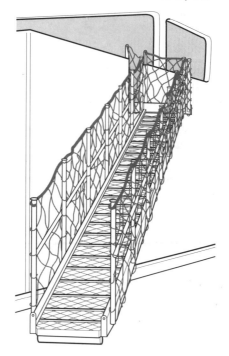

— it is safe to pass from the lowest tread or platform of the ladder onto the dock and also up to and onto the deck.

4. As far as is practicable, the ladder should be kept free of any snow, ice, grease or other substance likely to make a handhold or foothold insecure.

5. Precautions should be taken to prevent the suspension ropes of accommodation ladders from becoming slack;

this can result in the ladder falling violently along the ship's side if the ship lurches away from the quay as a result of wind or the wash of a passing ship.

7.2.3. Gangways

1. Gangways should be placed at right angles to the ship's side and connect the quay with one of the ship's decks or with the bottom platform of the accommodation ladder.

2. A gangway should:

— have a closely boarded walkway at least 550 mm in width;

— be fitted with transverse treads at suitable stepping intervals;

— be fitted with upper and intermediate guard rails;

— be fitted with devices enabling it to be properly secured to the ship;

— be fitted with proper slinging attachments so placed that it will balance about the attachments when it is suspended, if a lifting appliance has to be used to place it in position;

— not normally be used at an angle of more than 30° to the horizontal, or 45° if it is fitted with transverse treads every 500 mm or so;

— never be used at an angle of more than 45° to the horizontal;

— land on a clear and unobstructed area of the quay.

3. Where a gangway rests on a ship's bulwark, safe means of access should be provided between the ship's deck and the gangway.

4. A gangway that rests on a quay on rollers or wheels should be positioned in such a way that the rollers or wheels are on a reasonably level surface and not in the vicinity of any obstruction or hole that could restrict their free movement. The rollers or wheels should be fitted or guarded in such a way as to prevent a person's foot from being caught between them and the quay surface, leaving a minimum gap of 50 mm (figure 97).

5. A gangway should be securely fastened to the ship. One arrangement consists of placing the gangway above the deck or the platform and providing a fastening at the bot-

Figure 97.   Gangway foot

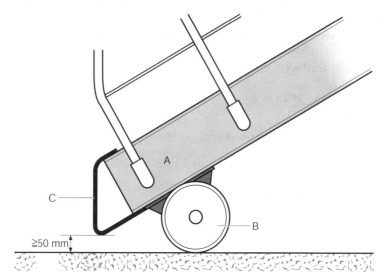

A. Bottom frame of gangway. B. Wheel or roller. C. Protective metal sheet.

tom, for example a U-section whose web, perpendicular to the gangway floor, is placed at the end of the gangway. The fastening should be supplemented by safety ropes or chains. Small chains should also be used to ensure the continuity of the handhold between the gangway railing and the ship's bulwarks.

7.2.4. Portable ladders

A portable ladder (see section 3.5.4) should only be used as means of access to ships in exceptional circumstances, such as in the event of damage to an accommodation ladder or a gangway.

7.2.5. Rope ladders

1. A rope ladder should only be used to provide access from a ship to a barge or similar vessel of lower freeboard (figure 98).

2. When a rope ladder is rigged:

— its two suspension ropes on either side should be under equal tension, as far as possible, and properly secured to the ship;

— the treads should be horizontal and constructed to prevent twisting;

— safe access should be provided between the top of the ladder and the deck of the ship and the lower part of the ladder and the other ship;

— the ladder should, when practicable, hang fully extended when in use and not be positioned over or in close proximity to a discharge opening in the ship's side.

Figure 98.   Rope ladder

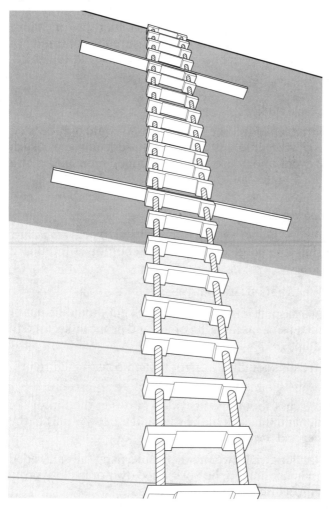

7.2.6. Bulwark steps

1. Bulwark steps are a type of stepladder placed on the ship's deck behind the bulwarks between the deck and a gangway ending at the level of the bulwark rails.

2. Handrails, or other firm handholds for users of bulwark steps, should be provided, if possible on both sides.

7.2.7. Access to the decks of bulk carriers and other large ships

Where the freeboard of the ship is too high for normal access equipment to be used and there is no lower access point in the ship's hull, specially designed equipment should be provided on the shore.

7.2.8. Access to barges and other small ships

Where the freeboard of the ship is too low to allow normal access equipment to be used, the ship should be moored in a position where the fixed ladders in the quay walls can be safely used for access.

7.2.9. Access to ro-ro ships

1. Whenever possible, safe pedestrian access, separate from vehicle access ramps, should be provided to ro-ro ships (figure 99A). Pedestrian access via the main loading ramp can present hazards from moving vehicles.

2. When pedestrian access via a vehicle ramp is necessary, a separate walkway or walkways should be provided on the outer edge or edges of the ramp (figure 99B). Walkways should be fenced on both sides to prevent falls into water and provide protection from moving vehicles.

349

3. If the provision of a fenced walkway is not practicable, a clearly marked and signed walkway should be provided on one side of the ramp (figure 99C).

4. If none of these options is practicable, access via the ramp should be controlled at all times while vehicles are using it (figure 99D). The degree of control that will be necessary may vary with the size of the ramp and the number of vehicle movements. The control arrangements adopted should ensure that all pedestrians, including seafarers and other persons visiting the ship, are subject to the same control system.

5. The ramp controller should ensure that when vehicles are using the ramp, pedestrians are prevented from doing so. The traffic movements should be stopped to enable them to transit the ramp. Control may be effected by hand signals or by traffic lights.

6. If the ramp or ramp and link span combination is long, it may be necessary to have a controller at both ends in direct radio communication with one another.

7. Controllers should wear high-visibility outer clothing and safety helmets at all times.

8. Portworkers driving trade vehicles to or from a ship and a terminal storage area should return by minibus or other suitable vehicle (figure 99E).

9. If loading or unloading of the ship takes place in tidal waters, a suitable link span or floating bridge should be installed, where necessary, to ensure that the slope of the ramp does not exceed 1 in 10. The ramp should be fenced to prevent a vehicle or a person from falling from its sides.

Figure 99.　Control of pedestrians on ramps

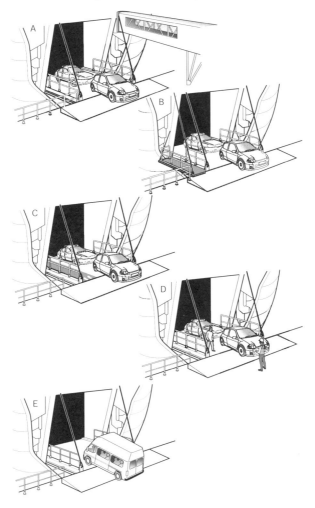

10. If a ramp is capable of dealing with simultaneous two-way traffic, or two separate ramps are in use, the directions of traffic should be clearly marked by arrows (figure 100). These should be clearly visible when vehicle movements take place at night or in poor light.

Figure 100. Traffic direction signs on a ship's ramp

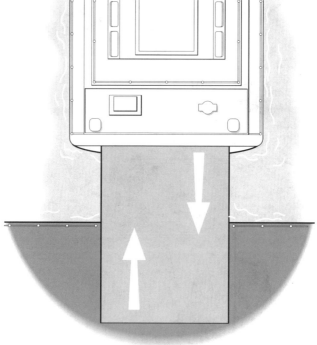

11. If a ramp is capable of carrying only one vehicle at a time, precedence on the ramp should be given to a loaded vehicle.

12. The edge of the quay adjacent to a ship's ramp should be protected to prevent people from falling into the water.

7.2.10. Ship-to-ship access

Where there is a need to gain direct access from one ship to another, gangways or other suitable access equipment should be used as appropriate.

7.2.11. Access by water

1. Access to ships by water can be hazardous and should be kept to the minimum. However, portworkers have to be transported by water when ships are worked away from a berth.

2. Tenders (see section 3.14) used to transport portworkers to or from ships should be fitted out for carrying passengers in compliance with national legal requirements.

3. At either end of the journey, portworkers should not board or leave the tender unless it is properly moored or secured. Particular care should be taken when passing between the tender and the ship. Wherever possible, access to the ship should be by its accommodation ladder.

7.3. Access on board ships

7.3.1. General requirements

1. Safe means of access about the ship should be provided for portworkers between the gangway or other main

access and the holds, deck cargoes, winches and cranes that are to be worked.

2. Access routes should not pass under cargo being worked.

3. All such access routes should be kept tidy and clear of obstructions. If specially constructed, they may consist of wooden grating or steel plates at least 600 mm wide. They should be raised some 100 mm above the deck.

4. Access routes should as far as possible avoid lashings, ropes and other obstructions that might otherwise impede access. If deck cargo is stowed up to the bulwarks, access should be provided on the other side of the ship or, if that is not possible, a safe route should be constructed through or over the cargo.

5. If access is required during the hours of darkness, the routes should be lit in accordance with section 7.1, paragraph 5.

6. Portworkers should always be alert to moving vehicles when moving around cargo holds and decks of ro-ro ships. Safety helmets and high-visibility garments should always be worn.

7.3.2. Access to holds

1. Access to cargo holds should be effected by the ship's permanent access (figure 101). Access should be effected by portable ladders only if all permanent access ways are obstructed or otherwise unable to be used.

2. Access hatches ("man-hatches") and other openings giving access to holds should be protected by coamings. There should be a clear space of at least 400 mm around the

Figure 101.   Hold ladder

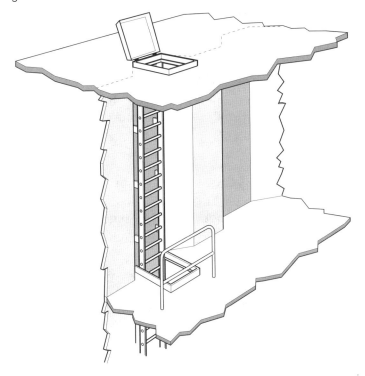

coamings to allow easy access. If openings are too small, coating the lower edges with foam rubber or other elastic material should ensure that heads and shoulders are not injured if they strike against them.

3. The approaches to a hold and an access hatch should be kept unobstructed to reduce the risk of falls and to enable holds to be evacuated quickly in an emergency.

4. Portworkers should be alert to the possibility that openings into holds have been left open or unfenced, or are hidden or obstructed by cargo.

5. Where such openings have lids, these should be secured to prevent them from accidentally closing during access.

7.3.3. Access to deck cargo

1. When it is necessary for portworkers to gain access to the tops of general break-bulk deck cargoes and safe means of access are not provided, suitable safe access should be constructed. This should include suitable handholds.

2. Access routes onto and about stows of timber on deck should be constructed, where necessary, in accordance with Chapter 5 of the IMO's *Code of Practice for Ships Carrying Timber Deck Cargoes*. When access involves walking across the cargo itself, care should be taken not to step into gaps in the timber stows and to avoid tripping hazards such as banding or pre-slung cargo slings lying on the surface of the timber. When uneven-length timber units have been wrapped, the wrapping on top should be removed.

3. Particular care should be taken when timber deck cargo is discharged, as rough weather may have dislodged stacks or made them unstable.

4. When possible, container top working and the need for access should be avoided. However, when container top

work is necessary on board ships, safe means of access should be provided (see section 7.8.3).

5. When there are no safer means available, portable ladders (see section 3.5.4) may be used to access containers up to two high. A co-worker should always be holding the ladder unless it is otherwise adequately secured. Port-workers should never be permitted to climb up the ends of containers.

7.4. Hatches

7.4.1. Hatch coverings

1. All hatch covers, hatch beams and pontoons should be:

— of sound construction and maintained in good condition;

— plainly marked to indicate the hatch, deck and section to which they belong, unless all such items are inter-changeable;

— fitted with effective locking devices that prevent them from being displaced when locked.

2. Hatch covers that have to be lifted by hand should be fitted with suitable hand grips. They should be inspected before use on each occasion. If any are found to be missing or defective, the necessary replacements or repairs should be made immediately by the ship's officers.

3. Wooden covers should be bound with steel bands or straps. The bands or straps should be kept firmly secured in place, particularly at their ends.

4. Broken, split, poorly fitting or otherwise defective covers and beams should not be used and should be immediately repaired or replaced.

5. The bearing surface of covers, pontoons and beams should be sufficient to support any loads which they have to carry and also be wide enough to practically eliminate any risk of a beam slipping out and falling into the hold. Seatings should be at least 65 mm wide for covers and 75 mm for beams.

6. If the hatch is fitted with rolling or sliding beams (figure 102):

— the top guide should project over the roller or end of the beam in such a way that when a beam is moved horizontally in the directions of its length as far as it will go in one direction, the other end is still retained by its top guide;

— lateral movement of a rolling or sliding beam in the horizontal direction of its length should be restricted in such a way that in the event of its crabbing it will seize before an end can slip off its guide;

Figure 102. General arrangement for a rolling/sliding hatch cover

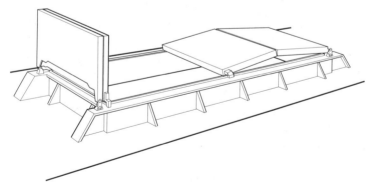

— it should not be used if components are missing, especially parts of the locking mechanism; deficiencies should be reported to a ship's officer;

— no attempt should be made to release a rolling or sliding beam that seizes in position while being moved by bumping it with a suspended load or by hauling on it by means of a wire rope reeved on the warping drum of a winch; the matter should be reported to a ship's officer.

7. Locking devices should be strong enough to withstand a reasonable blow from any swinging cargo without damage. The devices may be self-locking, so that the locked component can be released only by a manual operation.

8. Folding hatch covers should be fitted with locking devices, wheel chocks or other suitable means to prevent the covers from spontaneously folding back when they are released from their coaming seats (figure 103).

9. Every hatch beam and cover that has to be removed by the use of a lifting appliance should be fitted with suitable attachments for securing the lifting slings or spreader. The attachments on beams should be so positioned that it is not necessary for a person to go on a beam to secure the sling or slings.

10. When pontoons are lifted by four-legged slings, the slings should be long enough to reach easily the holes at the ends of the beam when forming an angle not exceeding 90° to each other.

11. Each leg of all beam and pontoon slings should be equipped with a substantial lanyard at least 3 m long.

Figure 103.   Hatch cover locking device

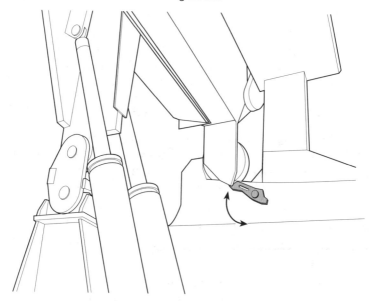

12. When cargo is to be stowed on deck or 'tween deck hatch covers, it is essential to ensure that the covers and their supports are of adequate strength to withstand the static and dynamic loads that they will have to bear.

13. Hatch covers, beams and pontoons should not be used in the construction of cargo stages or for any other purpose that might expose them to damage.

7.4.2. Handling hatch coverings

1. Hatch covers, beams and pontoons should not be removed or replaced while work is going on in the hold under the hatchway.

2. Portworkers removing hatch covers by hand should work from the centre towards the sides and from the sides towards the centre when the covers are being replaced. Workers should use suitable long-handled hooks to avoid stooping when grasping the covers or sling halyards. Hatch covers that cannot be easily handled by two workers should not be handled manually.

3. Portworkers pulling tarpaulins should walk forwards, where possible, not backwards, when working on hatch covers.

4. Power-operated hatch covers should only be operated by designated members of the ship's crew or other authorized persons.

5. It is essential to ensure that there are no loose objects on folding or lift-away hatch covers before they are operated.

6. No person should be permitted to be on any hatch cover, whether closed or retracted, when it is about to be opened or closed.

7. Persons should be warned, by warning devices or otherwise, when hatch covers are about to be opened or closed.

8. Portworkers should keep well clear of hatch covers and their machinery while they are being operated. They should never stand on covers during such operations.

9. No person should be allowed onto the top of a retracted back-folding hatch cover unless the preventer chains or other securing devices are in position.

10. Loading or unloading should not take place at a hatch unless:

— all parts of a hatch covering that may be displaced by a load have been removed or secured;

— powered hatch covers are secured in the open position, or are of such a design as to make inadvertent abrupt closing impossible.

7.4.3. Stacking and securing of hatch coverings

1. Hatch covers, beams, tarpaulins and pontoons that have been removed should be placed, stacked or secured in such a way that they cannot fall into the hold, present a tripping hazard or otherwise cause danger.

2. Hatch covers and pontoons should either be arranged in neat stacks not higher than the coaming and away from it, or be spread one high between coaming and rail with no space between them. It is recommended that on the working side of the hatch the top level of the stacks should be at least 150 mm below the top of the coaming.

3. Hatch beams should be laid on their sides or stood on edge close together. They should be lashed, to prevent the outside one being overturned, and, if necessary, should be wedged to prevent tilting. Beams that are convex underneath should be wedged at each end.

4. The height of stacks should be limited so that workers below or overside will not be endangered if the stack is accidentally struck by a load.

5. A space of at least 1 m should be left between hatch covers, beams and pontoons that have been removed and the hatchway, if the construction of the ship so allows. If this is not possible, the covers should be stacked on only one side of the hatch and the other sides left free. Safe walkways should

also be left between the hatch coaming and the rail and from fore to aft.

6. The handling of pontoon hatch covers requires extreme care. The stacking guidelines in operation manuals and markings on hatch covers should be followed. All turnbuckles and lashing rods lying on the pontoon should be secured and any that hang over the edge removed.

7.4.4. Protection of hatches

1. Hatches at deck level should be protected by coamings of sufficient height to prevent accidental falls into the hold. Coamings should preferably be 1 m high.

2. Portworkers should not work on cargoes on deck or between decks that are over an opened hatch.

3. Work in the 'tween decks area should not normally take place if the hatch is open to the lower hold. If this cannot be avoided, the hatch should be fenced to prevent portworkers from falling.

4. The fencing should be 1 m high and may be of suitable wire rope or chain (figure 104(1)), provided that:

— there are means to keep the ropes or chains as taut as possible;

— wire ropes have sufficient wires per strand to be flexible, are free from broken wires, and any loose ends are fitted with ferrules or other means of protection to prevent injury;

— sufficient stanchions are provided.

5. Deck sockets into which stanchions fit should be equipped with locking devices and should be sufficiently deep

363

Figure 104.   Moveable fencing for 'tween decks (locking device
omitted for clarity)

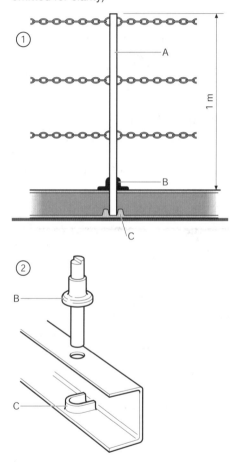

1. General view. 2. Detail.
A. Stanchion (steel tubing). B. Steel collar welded to the stanchion. C. Welded round iron.

and designed in such a way as to prevent the stanchions from moving unduly out of the vertical or being accidentally displaced (figure 104(2)).

6. The fencing should form a permanent part of the ship's equipment and be kept in place at all times, except:

— when the hatch is being opened or closed;

— when goods are being loaded onto that particular deck and the work in the hold prevents the hatch from being closed;

— during meal breaks or similar short interruptions of work.

7. Where necessary, barriers should be installed to prevent:

— lift trucks working in the 'tween decks area from falling into the hatch;

— lift trucks operating to and from side doors from falling onto the quay.

7.5. Work in holds

7.5.1. General requirements

1. The possibility that the atmosphere in a hold or accessway to a hold may be hazardous (see sections 9.1.7 and 9.2) should always be considered before entry is made.

2. The main hazards that portworkers handling cargo in holds should be aware of include:

— falling through openings in holds or from cargo;

— falls of unstable cargo;

— congested working areas;

— uneven working surfaces on cargo;

— tripping hazards;
— manual handling hazards;
— unclear or inadequate communication with lifting appliance operators;
— swinging loads;
— falling objects;
— mechanized plant and vehicles, and their fumes.

7.5.2. Working practices

1. The number of portworkers in each gang working in a hold with the same lifting appliance should depend on the nature of the cargo, the hours of work, the equipment used, the target output, and the fatigue that would result if not enough workers were employed. Numbers should be limited to what is necessary, since too many workers may be a hazard in the often constricted working spaces in a hold.

2. All persons working in holds should wear safety helmets and high-visibility clothing.

3. Goods should be stowed, handled, stacked or unstacked under competent supervision.

4. As far as is practicable, only one gang at a time should work in a hold. Where two or more gangs work in the same hatch:

— there should be a separate signaller for each fall worked, except in the case of union purchase;
— where gangs are working at different levels, a net should be rigged and securely fastened to prevent cargo from falling onto portworkers below.

5. When it is impossible for lifting appliance operators to have a clear view of the area where loads are being slung

in the hold, a signaller should be employed to direct the operator of the lifting appliance. The signaller should be able to see and be seen both by the portworkers in the hold and by the lifting appliance operator. The signaller should use an agreed set of hand signals (see section 5.4, paragraph 6). Alternatively, the signaller should be in direct radio communication with the lifting appliance operator.

6. Safe access to a safe position on the deck or the deck cargo should be provided for the signaller.

7. As far as is practicable, sets in holds should be made up in such a way that they can be lifted vertically. Lifting appliance operators should ensure that there is a smooth take up of the load and when it is lifted out of a hold. Where practicable, tag lines may be used to control any swing or twist motion.

8. Portworkers should stand away from the set once it has been made up and while it is lifted out of a hold. They should be alert to possible swinging of the load once the appliance has taken the strain.

9. When cargo is built up in sections in the hold, each section should allow for a safe landing place for the cargo.

10. Suitable protection should be provided where portworkers have to work close to edges from which they can fall more than 2 m. This may take the form of netting or other suitable means.

11. No loose gear or other objects should be thrown into or out of the holds.

12. Dunnage should be used when necessary to make cargo stowage safe and stable. When dunnage is used for this or any other reason, consideration should be given to

how it can be removed at the port of discharge and at any intermediate ports where access may be required.

13. Consideration should be given to the order of discharge when cargo is stowed between decks to ensure that when 'tween deck hatch covers and beams have to be removed, there will be a working space 1 m wide between the stowed cargo and the coaming. No such space needs to be left free on the covered portion of a partly opened hatch, but measures should be taken to prevent stowed cargo from falling into the open section.

14. Where cargo for discharge is situated under the 'tween decks area, it needs to be brought out to the square of the hatch where it can be plumbed by the lifting appliance in order to be discharged safely. Light goods can be moved on rollers into the square. With heavier goods, a suitable lift truck or other mechanical device should be used whenever possible. Where there is no alternative and the weight of the cargo is within the safe working load of the lifting appliance, a bull wire may be attached to the goods and reeved through a sheave on the opposite end of the hatch (figure 105). The bull wire should preferably be attached to a ship's winch. If a crane has to be used, the cargo hook should be attached to a bull wire to prevent wear on the hoist rope. The crane jib head should be positioned vertically above the sheave. One signaller should be on deck and another signaller in the hold to ensure that the goods do not snag. With careful movements, the crane should be able to bring the goods to the square.

15. Mechanical plant used in a ship's hold should:
— be fitted with an overhead guard;

Figure 105.   Use of bull wire

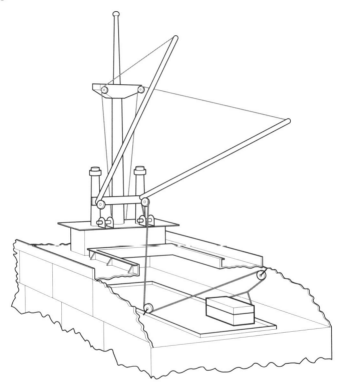

— have its wheels situated within its main body;

— be painted in a highly visible colour; if it is a rear-wheeled steered truck, the back portion should be painted with stripes or some other highly visible marking;

— preferably be electric or diesel powered;

369

— preferably all have the same or similar methods of control.

16. Mechanical plant that needs to be lifted in and out of holds should have:

— four lifting lugs built onto the body of the plant as near to the four corners as possible;

— each of the pair of lugs at the front and rear of the plant at the same height, but not necessarily at the same height front and rear;

— the lugs positioned so that when a sling is attached, its legs will not come into contact with the overhead guard or any other part of the plant.

17. A dedicated four-legged lifting sling assembly should be made up and used to lift each item of plant or similar items having the same layout, weight and configuration of lifting points (figure 106). The assembly should:

— have legs constructed from an appropriate size of steel wire rope;

— have legs of sufficient length to ensure that the plant remains level when lifted;

— preferably include a small chandelier spreader;

— be attached to the plant by shackles that form part of the assembly;

— be clearly marked with the identity of the item of plant or model of plant for which it is intended to be used;

— only be used for its intended purpose;

— be kept in the gear store away from all general-purpose slings when not required for use.

18. When mechanical plant is used in a hold, adequate ventilation should be maintained at all times.

Figure 106. Lifting frame and attachment points for lift truck (all the proper guards have been omitted for clarity)

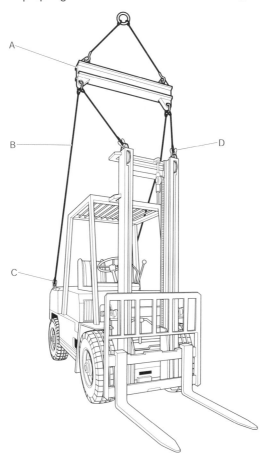

A. Spreader beam. B. Wire rope slings. C. Shackles attached to lifting lug of truck body.
D. Shackles attached to lugs at top of mast.

19. The tank top or the 'tween decks surface on which the plant operates should be of sufficient strength to support the weight of the plant and fully level. If the plant cannot operate safely on the floor of the hold, it may be necessary to put down steel plates or other temporary flooring to provide a suitable level and safe working surface.

20. When work is carried out in the 'tween decks area, the height of the mast at full lift should be restricted. Trucks with low masts should preferably be used.

21. The operator should pay careful attention to the stability of the plant at all times.

22. Working space in holds is often congested. Rear-wheeled steered plant has a tight turning circle. Great care should be taken by plant operators and other workers in the hold to avoid collisions with people, cargo stacks, which could be dislodged, or damage to the ship's structure.

7.6. Work on deck

7.6.1. General requirements

1. All upper decks to which portworkers may have access to carry out port work should be provided on the outer edge with a bulwark or guard rails that are so designed, constructed and placed, and of such a height above the deck, as to prevent any worker from accidentally falling overboard.

2. The bulwark or guard rails should be securely fastened in position. Removable sections should be securely fastened when in position.

3. Work surfaces should be safe, with ropes, beams and covers, hatch lids and other gear and equipment stowed safely

and tidily. Any spillages of oil, cargo or other substances likely to affect the safety of portworkers should be cleaned up. Portworkers should be alert to possible obstructions built into the deck, such as eye bolts, and those on the deck, such as lashings.

4. Deck cargoes should be stowed in such a way that:

— safe access is provided to the deck cargo, winches/deck cranes, hold ladders and signaller's stand; where necessary, properly secured ladders or other means should be used;

— winches and deck cranes to be used during loading or unloading can be safely operated.

5. When a signaller has to move from the square of the hatch to the ship's side, a space at least 1 m wide should be kept clear.

6. If the surface of the deck cargo is uneven, safe walkways running both fore and aft and athwartships should be provided, where this is practicable.

7. When deck cargo that is not being worked is stowed against the ship's rails or hatch coamings, and at such a height that the rails or coamings would not prevent portworkers from falling overboard or into the open hold, temporary fencing should be provided.

8. When cargo is stowed on deck or in the 'tween decks, and the hatches have to be opened at intermediate ports before that cargo is unloaded, it should be stowed in such a way as to provide a clear space of at least 1 m around the coamings or around the part of the hatch that is to be opened at subsequent ports.

9. If this is impracticable, suitable fencing should be rigged to enable portworkers to remove and replace in safety all fore and aft beams and thwartship beams used for hatch covering, and all hatch coverings.

10. The coaming clearance of 1 m should be marked with a painted line.

11. If goods have to be stowed on hatch covers, due regard should be paid to the bearing capacity of the covers. The responsible persons should satisfy themselves that:

— the hatch beams are properly placed; and

— the hatch covers sit well and fit tight together, and are in an undamaged condition.

12. When deck cargo is carried, adequate arrangements should be made to ensure that the signaller can be seen by the crane or winch operator. The signaller should have an unobstructed view of the hold and the winch operator.

13. When forest products, usually timber, have been carried on deck:

— allowance should be made for any weight gain due to water absorption;

— great care should be taken before the deck cargo lashings are let go; all personnel should be kept clear, in case there has been movement of the cargo that has put the lashings under tension and the cargo has become unstable.

7.6.2. Lashing and securing of cargo

1. All ships of 500 gross tonnes or more that are designed to carry cargo that requires lashing and securing for sea voyages are required to carry a cargo securing manual. This should detail how the cargo should be secured, what

lashings or other equipment may be used and how tight the lashings should be.

2. The stevedore should ensure that the requirements of the manual are followed, unless otherwise instructed by the master of the ship. General guidance on the securing of cargo is contained in the IMO's *Code of Safe Practice for Cargo Stowage and Securing (CSS Code)* and the Nautical Institute's book *Lashing and securing of deck cargoes*.

3. Safe places of work should be provided to enable portworkers to carry out such securing work.

7.7. Shot cargo

1. Ships at sea can often encounter bad weather and rough seas. This can cause cargo to move, despite all the securing arrangements made beforehand. Deck cargoes can be moved out of stowage and even lost overside. Hold stowages can move and, in severe cases, be completely mixed up and out of place. In such circumstances, damage may be done to cargo and spillage may occur, including spillage that continues after the adverse conditions have passed.

2. Ships that have encountered bad weather and rough seas may come into port experiencing difficulties with their cargo. If the ship itself has a problem, that should be dealt with first. When the ship is safe and is at its berth, the stevedore should stabilize the cargo before discharging it or securing it for a further voyage.

3. Great care needs to be taken to ensure the safety of portworkers during such operations, which should normally be carried out under the direct control of experienced supervisors. A high level of alertness is required, particular

attention being paid to the stability of the cargo, safe access, footholds and handholds, the application of lifting gear and the need to stand well clear.

7.8. Container ships

7.8.1. General requirements

1. Containers stowed in open hatches that are secured by the cell guides do not need further securing arrangements.

2. Containers carried by ships that do not have cell guides should be secured by lashings, bars, etc., both in the hold and on deck.

3. All lashing gear is provided by the ship and is ship's equipment. Fully manually operated twistlocks are now tending to be replaced by semi-automatic twistlocks (SATLs). On loading, SATLs may be placed in position on the underside of the container on the quay. When the crane lowers the container into position, the SATLs automatically lock into position. On discharge, the SATLs have to be unlocked with the aid of a long pole. Such poles can only be used from deck level to unlock up to four containers high because of their length and weight.

4. The operators of container quay cranes should be positioned in such a way that they can see directly down onto the cargo and the crane, and thus lock on to individual containers and lift them without other persons being involved.

5. The need for working on top of containers should be eliminated or reduced by the use of:
— hatchless ships that eliminate it;

— SATLs that reduce the need but do not eliminate it completely;

— a combination of lashing platforms and SATLs restricting it further;

— fully automatic twistlocks.

6. When a jib crane or derrick is used for discharging/ loading, there may be a need to steady the load when a container is being lifted or lowered, or a spreader is lowered onto a container.

7. When it is necessary to use over-height frames to lift open-topped containers:

— all lifting brackets, shackles and other loose gear on both the main frame and the subframe used in the lift should be subject to the provisions of section 4.3 and marked accordingly;

— the over-height frame should be marked with its weight and safe working load;

— a physical check that twistlocks have turned and are engaged should be made before lifting;

— where necessary, care should be taken to ensure that the twistlock operating ropes do not catch on fixed objects while the frame is in use;

— frames should be securely stowed on trailers when not in use.

8. When container cranes are used to lift loads other than freight containers, it should be ensured that:

— the equipment and methods are adequate and safe;

— the manufacturer's recommendations are followed if the crane's heavy lift hook is used;

— lifting frames are not asymmetrically loaded beyond the manufacturer's recommendations;

— only tested and marked lifting points on the main frame or other frames are used.

9. Further guidance on the safe operation of container cranes and work with them is included in Chapters 5 and 6.

10. Further general guidance on safe work on container ships is contained in the ICHCA International Ltd. *Safe working on container ships,* International Safety Panel Briefing Pamphlet No. 8.

7.8.2. Deck working

1. Shore-side management should ensure that safe access is provided by the ship to any place on the ship where stevedores have to work, and that the place of work is safe.

2. The placing and removal of lashing equipment on the ends of containers should be carried out in the athwartship gaps between container stows.

3. The space provided between the container stows for portworkers to carry out such work (figure 107) should provide:

— a firm and level working surface;

— a working area, excluding lashings in place, preferably of 1 m and not less than 750 mm wide to allow clear sight of twistlock handles and the manipulation of lashing gear;

— sufficient space to permit the lashing gear and other equipment to be stowed without causing a tripping hazard.

Figure 107.   Working space for placing and removing lashing
equipment

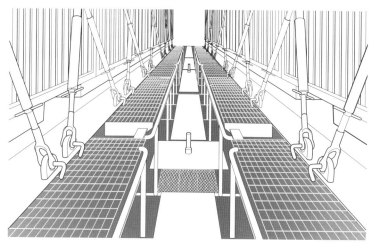

7.8.3. Container top working

1. When work on container tops cannot be avoided,
safe means of access to them should be provided.

2. Access to the tops of containers should be from part
of the ship's permanent superstructure whenever possible.
This may be from lashing platforms.

3. When such access is not possible, [2] safe access should
be provided by the use of a quayside crane and a:

— purpose-built access cage (see section 3.6.2.9);

— purpose-built gondola;

---

[2] Some countries do not allow riding on platforms.

— platform built on the container spreader;
— platform built on the headblock.

4. When a cage or platform is used for access:

— at least two persons should travel in the cage or platform, one of whom should have a radio in direct contact with the crane operator;
— the crane operator should only take directions from that person;
— the secondary means of attachment to the spreader should be connected;
— all parts of the body, particularly the hands and head, should be kept inside the cage or platform at all times.

5. Portworkers should never climb up the ends of containers.

6. Non-purpose-built container-carrying ships may also carry containers on deck or in the hold in circumstances where portworkers may be required to access container tops. When this involves loading or discharging by jib crane, an additional reason for being on the top layer of the containers may be to steady the load as it is positioned or removed. In these circumstances, a safe system of work should be developed to ensure that portworkers have safe access.

7. When work has to be undertaken on container tops, precautions should be taken to ensure the safety of portworkers. Suitable fall prevention or fall arrest systems of work should be devised and used in order to eliminate or control the risk of falling from the container stow. Fall prevention systems include working from inside a cage used for access, or secured to a short lanyard that prevents falls from open sides of containers.

8. The choice of system actually used will be influenced by the equipment used to secure the containers. If this equipment consists of manually placed twistlocks and bridging pieces, it may be possible to carry out the work from inside an access cage, or it may need to be undertaken actually on the tops of containers. If the securing equipment consists of SATLs, there should be no need to go onto container tops during loading operations. On discharge, SATLs above four containers high have to be unlocked by pole either from the topmost tier or from a gondola at the side of the stow.

9. When a purpose-built access cage is used, it can be moved slowly across the top of each tier of containers while workers in it place or remove twistlocks. Great care should be taken to ensure that portworkers' hands are not trapped. A second person in the cage should be in direct radio contact with the crane operator and should control the operation at all times.

10. When it is necessary for portworkers to leave an access cage or platform to go to the corners of the containers, carrying the twistlocks, bridging pieces or locking poles, etc., with them, they should wear a full body harness and be connected to a secure anchorage point by lanyards, safety lines or inertia reel fall arrest equipment. The harness should have "D" rings at the front and back for attachment to the reel and to aid recovery.

11. Other systems or methods may be used in connection with container top working, provided that they ensure the safety of portworkers at all times.

12. Work on top of containers should cease in high-wind conditions (see section 11.1.9).

13. Similar precautions should be taken to ensure the safety of portworkers who have to go onto the tops of containers on the deck or in the hold of combi ships, where freight containers are stowed and lashed.

14. Further guidance on safe working on tops of containers is in the ICHCA International Ltd. *Container top safety, lashing and other related matters,* International Safety Panel Research Paper No. 4.

7.9. Ro-ro ships

7.9.1. General requirements

1. Ro-ro ships are equipped with a variety of cargo access equipment, e.g. ship/shore ramps, bow/stern/side doors, internal ramps and cargo lifts. This equipment is normally operated by the ship's crew.

2. The main operations in a ro-ro hold are the marshalling of vehicles and lashing them to the deck for the voyage. In a sto-ro ship, cargo such as paper reels is brought into the hold on roll trailers. It is then taken off the trailer by lift truck and placed into a stow in the hold.

3. In each of these operations, mechanical appliances are widely used and, apart from driver-accompanied freight vehicles and passenger cars, are usually driven and operated by portworkers, who may also marshal vehicles and lash/unlash vehicles to the deck.

4. The principal hazards for portworkers working in ro-ro holds are associated with vehicle movements. Vehicles moving in a confined space represent a risk of person/machine contact and vehicle exhaust fumes can affect

health. Lashing operations can also present a risk. Portworkers should also be aware of any cargo-access equipment in the area where they are working and know how it operates.

5. Audible and visual warnings should be given before any cargo-access equipment is operated. Portworkers should be alert for such indications.

6. The slope of an internal ramp should not exceed 1 in 10.

7. Every stanchion or other fixed structure on an enclosed deck that is liable to be a danger to vehicles, or to give rise to a risk of trapping between itself and a vehicle, should be clearly marked with alternating black and yellow stripes.

8. All portworkers on ro-ro ships should wear high-visibility clothing.

9. For access to ro-ro holds, see section 7.2.9 and for precautions against vehicle fumes, see section 9.1.7.

7.9.2. Vehicle movements

1. All movement of vehicles on board ro-ro ships should be effectively and continuously controlled.

2. Only authorized persons should be allowed on any vehicle deck while vehicle movements are taking place.

3. Drivers should comply with the relevant speed limits on ramps and vehicle decks at all times. These may be lower than those on the quay. Signs indicating the speed limit should be clearly displayed in prominent positions both on the quay and on the ship.

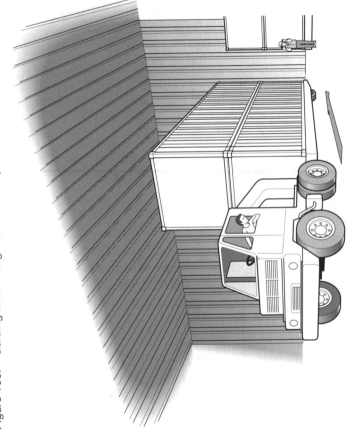

Figure 108. Guiding a reversing trailer on a ship's deck

4. All large vehicles and trailers being reversed or manoeuvred into stowage positions on deck should do so under the direction of a signaller (figure 108). Signallers should satisfy themselves that no person is in a position of danger, particularly in any trapping area behind a reversing vehicle. Drivers should not move their load/vehicle unless a signaller so directs. Drivers should immediately stop their vehicles at any time the signaller is not within their field of vision.

5. While loading and unloading is taking place, the area should be kept clear, so far as is practicable, of dunnage, loose wires, unused vehicles, securing gear and other extraneous equipment or material.

7.9.3. Passenger movements

1. The movement of passengers on foot on the vehicle decks of ro-ro ships should be strictly controlled and minimized.

2. Passengers arriving on ships in vehicles should:
— be given clear instructions for parking their vehicles;
— apply the handbrake before leaving their vehicles;
— be prohibited from walking around vehicle decks while loading is taking place except on clearly marked safe walkways.

3. Passengers returning to vehicle decks prior to discharge should:
— only enter the vehicle deck when permitted and by an authorized route;
— not be permitted to start engines until ramp doors open;

— not be permitted to move until the vehicle deck is clear of passengers;

— not be allowed onto the vehicle decks after vehicles have started to move except on clearly marked safe walkways.

7.9.4. Vehicle lashing operations

1. The wearing of bump caps by portworkers lashing vehicles may be more appropriate than safety helmets owing to the restricted working positions.

2. Portworkers carrying out lashing operations should work in pairs, each worker always remaining in sight of the other.

3. Great care should be taken when vehicles are moving, especially when the system requires vehicles to reverse into place. In particular, it is essential to ensure that:

— large vehicles are always controlled by a signaller when reversing (see section 7.9.2);

— portworkers do not position themselves at the back of a vehicle when vehicle loading operations are taking place in that row.

4. Portworkers should release lashings warily, as vessel and vehicle movement during the voyage may have made them excessively taut.

7.9.5. Cargo lifts

Portworkers working on or near cargo lifts should:

— not ride on a cargo lift when it is in operation, except the driver of a vehicle who is in the cab;

— exercise caution when working with or close to a cargo lift.

7.10. Bulk carriers

1. Loading and unloading should be undertaken in accordance with the plan required by the IMO *BLU Code* (see section 6.19) and agreed between the terminal representative and ship's master.

2. When portworkers are required to use mechanical plant in a hold to trim loaded dry bulk cargo and to move or break down cargo residues for discharge, care should be taken not to cause damage to the structure of the ship.

3. During loading, the regular distribution of the cargo in holds should be achieved by careful use of the loading machinery. This may be supplemented by belt conveyors or rotary machines, which throw the material some metres from the loading spout.

4. During unloading, cargo may have to be moved into an area of the hold from where it can be picked up by grabs or by other handling equipment such as suction pipes and pneumatic equipment. This may be carried out with mechanical plant, such as vehicles fitted with a bucket or, in some instances, by hand.

5. When work has to be carried out in the holds of bulk carriers:

— a signaller may be necessary to control grabs or other equipment;

— in holds loaded by grabs, one worker should act as look-out if there is a danger of workers being buried under a load from a grab;

— all trimmers should be checked in and out of the hold;

— workers should be secured by a full safety harness and lifeline when necessary during trimming or discharge;

— appropriate precautions should be taken to prevent dust inhalation;

— the equipment and methods used to bring down bulk cargo residues adhering to the sides and ends of holds should ensure the safety of workers;

— equipment such as grabs should only be used for the purpose for which it was designed.

6. When equipment is being used in a bulk cargo hatch, no person should work unobserved.

7.11. Hot work

1. In the case of hot work in or near tanks, a gas-free certificate issued by a chemist or other suitably qualified person approved by the authority should be obtained. The certificate should be renewed if circumstances change, and in any case at least every 24 hours.

2. In special cases, such as hot work in or near the holds of tankers or combination carriers, a thorough inspection of the area should be conducted by specialists who can determine whether specific safety measures are required. Pipe work and pumps on board ships that have carried flammable liquids or gases should also be certified as gas free.

8. Dangerous goods

8.1. Packaged dangerous goods

8.1.1. General requirements

Many cargoes transported in packages have hazardous properties that could cause fire and explosion, injuries or environmental damage. Emergencies could occur anywhere in the transportation chain. However, as a result of internationally recognized rules for carriage by sea, which have applied since 1965, millions of tonnes of dangerous goods are safely handled at ports every year.

8.1.2. International arrangements

The transport by sea of dangerous goods in packaged form is required to be carried out in accordance with the *International Maritime Dangerous Goods Code (IMDG Code)*. The Code became mandatory from 1 January 2004 under the provisions of Chapter VII of the IMO's SOLAS Convention. Produced by the IMO, it is based upon recommendations published by the United Nations Committee of Experts on the Transport of Dangerous Goods. The *IMDG Code* is revised and republished every two years.

8.1.3. United Nations classification

1. The United Nations (UN) system of classification of packaged dangerous goods is an integral part of the international provisions. Goods to be transported are classified according to their primary hazard by the shipper or consignor. The nine UN classes are:

— class 1 – explosives;
  • subdivided into six divisions 1.1-1.6;

— class 2 – gases;
  - subdivided into flammable, non-flammable and toxic gases;
— class 3 – flammable liquids;
— class 4 – solids;
  - subdivided into flammable, spontaneously combustible and dangerous when wet;
— class 5 – substances containing oxygen;
  - subdivided into oxidizing agents and organic peroxides;
— class 6 – toxic substances;
  - subdivided into toxic and infectious substances;
— class 7 – radioactive substances;
  - subdivided into three separate levels of radioactivity plus fissile material;
— class 8 – corrosives;
— class 9 – miscellaneous dangerous goods not covered by the other classes.

2. The *IMDG Code* also recognizes that many substances, as well as being potentially dangerous to humans, can be environmentally hazardous to the marine environment. Accordingly, for the marine mode only, it uses the term "marine pollutants" for those dangerous goods to which this applies. There are also two UN Numbers in class 9 for solid and liquid substances that are not hazardous to humans but are hazardous in the marine environment, e.g. creosote.

8.1.4. *IMDG Code*

1. The *IMDG Code* contains the international require-ments for the safe transport of dangerous goods by sea. This Code is mandatory for all IMO Member States. This means that their national legal requirements as flag States require dangerous packaged goods transported by sea in ships flying their national flag to be carried in accordance with the Code. As the ILO Occupational Safety and Health (Dock Work) Convention, 1979 (No. 152), refers to dangerous goods on the shore side, the combined legal requirements should cover the whole journey from the entry to the export port to the destination in the import port.

2. Additional legal requirements may apply to move-ments by road, rail or air.

3. The basic requirements of the *IMDG Code* are that all packaged dangerous goods are:
— classified in accordance with the UN system of classification;
— packaged in appropriate UN approved packagings;
— labelled;
— packed onto a cargo transport unit, when appropriate;
— declared.

4. The shipper is required to classify the goods and arrange for them to be packaged in appropriate UN ap-proved packagings. The resultant packages are required to be labelled with the relevant hazard warning signs.

5. Individual packages are often further packed inside a cargo transport unit (CTU), such as a container or road or

rail vehicle, in which it will be transported along the transportation chain. The dangerous goods in the CTU should be segregated, packed and secured in accordance with the IMO/ILO/UN ECE *Guidelines for Packing Cargo Transport Units (CTUs)*, with the relevant hazard warning signs affixed to the outside of the CTU and a container or vehicle packing certificate, certifying the correct packing of the goods and other matters, completed and signed.

6. The hazard warning signs may be labels, placards, marks or signs. These are essential to alert personnel throughout transportation to the presence and hazards of the dangerous goods. Labels are affixed to packages and placards to CTUs. The *IMDG Code* specifies how many labels/placards are to be attached (figure 109).

7. Diamond-shaped labels and placards identify hazards by colour and symbol. The design for each class is different and, for classes 2, 4, 6 and 7, there are also different designs for the subdivisions. The class number, and for classes 1 (divisions 1.1, 1.2 and 1.3 only) and 5, the subdivision is shown at the bottom of the label and placard. For class 1 the compatibility group is also shown. Appropriate diamonds are required to be affixed for the primary hazard and up to two other subsidiary hazards.

8. The marine pollutant mark, elevated temperature mark and fumigation warning sign are also required to be affixed when relevant. In addition, packages and certain CTUs are required to be marked with the Proper Shipping Name and the UN Number.

9. All packaged dangerous goods for transport by sea are required to be declared in a dangerous goods transport

document signed on behalf of the shipper. When relevant, this should be included or be accompanied by a container/vehicle packing certificate.

10. The information to be included in the declaration always includes:

— Proper Shipping Name;

— class and where necessary division;

— UN Number;

— packing group;

— number and kind of packages;

— total quantity of dangerous goods.

11. Information that may also be required includes:

— the words "marine pollutant" where applicable;

— the words "limited quantity" where applicable;

— special information for goods in classes 1, 6.2 and 7, for certain substances in classes 4.1 and 5.2, and for CTUs under fumigation;

— minimum flashpoint, if 61ºC or less;

— specific reference to empty, uncleaned packages, portable tanks and bulk packagings and waste dangerous goods;

— subsidiary hazards not conveyed by the Proper Shipping Name;

— other information required by national authorities;

— weathering certificates, exemption certificates and classification certificates for certain substances in classes 4.1 and 5.2.

Figure 109. *IMDG Code* hazard warning labels, placards, marks

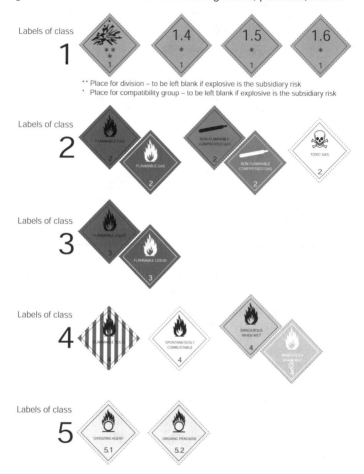

Labels of class **1**

** Place for division – to be left blank if explosive is the subsidiary risk
* Place for compatibility group – to be left blank if explosive is the subsidiary risk

Labels of class **2**

Labels of class **3**

Labels of class **4**

Labels of class **5**

Labels of class 6

Labels of class 7

Labels of class 8

Key:
- Orange
- Red
- Green
- Blue
- Yellow

Labels of class 9

MARINE POLLUTANT
Mark

ELEVATED TEMPERATURE
Mark

FUMIGATION WARNING
Sign

DANGER

THIS UNIT IS UNDER FUMIGATION
WITH [fumigant name*] APPLIED ON
[date*   ]
[time*   ]

DO NOT ENTER

* Insert details as appropriate

For further information on the use of labels, marks and signs, see Part 5 of the *IMDG Code.*
Source: Reproduced with the permission of the International Maritime Organization.

12. The information is required in order that the port and shipping company can arrange for safe handling, stowage and segregation on the terminal and on board the ship. Shipments should not be allowed to proceed into the maritime transportation chain without such information being properly provided.

13. Some substances, if allowed to come into contact with one another, will react and may cause a serious incident. Consequently, they need to be segregated both within CTUs and on board ships. The segregation requirements in the *IMDG Code* require segregation on ships both on deck and under deck. There are four segregation distances: 3, 6, 12 and 24 m. In some instances vertical separation is also required. The same provisions can be used for segregation on the terminal.

14. Other requirements relate to the carriage of small amounts of dangerous goods, known as limited quantities. Certain of the requirements are relaxed for small amounts contained inside receptacles and packages. These relaxations only apply to the less dangerous goods.

15. While the *IMDG Code* is intended mainly for precautions to be taken for the sea voyage, its provisions can also be applied in shore-side terminals and the Code recommends that it be so used.

8.1.5. Checking of packaged dangerous goods

1. On receipt of the documentation, checks should be made on the accuracy of the contents. This should include the information in 8.1.4.10 and that the Proper Shipping Name and UN Number are consistent. The container or vehicle packing certificate should be completed, where rele-

vant, and both the certificate and the declaration should have been signed.

2. The procedures should ensure that those concerned know what action should be taken in the event that the information is not fully correct.

3. When the vehicle arrives at the port or terminal, a check should be made that the placards, signs and marks are affixed to the outside of the vehicle or container in the required numbers and positions.

8.2. Solid bulk cargoes

1. Some solid bulk cargoes can be hazardous when shipped in bulk. The most common bulk cargoes are coal, metal ores, agricultural cargoes including grain, fertilizers and fertilizer raw material, and scrap metal.

2. Hazards associated with the transportation of solid bulk cargoes include:
— their inherent dangerous properties, covered by the nine UN classes (see section 8.1.3);
— other relevant properties;
— operational hazards.

3. Relevant considerations may include:
— oxidation resulting in lack of oxygen in a hold, accessway or other confined space;
— decomposition resulting in evolution of toxic or flammable gases and possibly also lack of oxygen;
— the angle of repose at which the cargo is likely to be unstable;
— their potential to liquefy;

— physical properties allowing cargo to collapse or persons to sink into it.

4. Operational hazards include:

— incorrect procedures;

— misdeclaration of cargo. Bulk shipping names should be used in accordance with the IMO's *Code of Practice for Solid Bulk Cargoes (the BC Code)*;

— lack of communication;

— unexpected presence of fumigants.

5. The IMO's *BC Code* lists the solid bulk cargoes carried by sea. Appendices A and B of the Code list those cargoes that may liquefy and those that have hazardous properties. Other cargoes that are typically carried by sea are listed in Appendix C of the Code.

6. Solid bulk cargo-handling operations should be carried out in accordance with the IMO's *Code of Practice for the Safe Loading and Unloading of Bulk Carriers (BLU Code)* (see section 6.19).

8.3. Bulk liquids and gases

1. Many bulk liquids and gases have hazards associated with their inherent chemical properties. In particular, many have low flashpoints and represent serious potential fire and explosion hazards. In view of the inherent risks and the volumes of cargoes stored and handled, such bulk liquids and gases should generally be handled at purpose-built terminals situated away from the main port facilities.

2. Bulk liquids and gases should be handled in accordance with the procedures set out in internationally recognized guidelines, including:

— IAPH/ICS/OCIMF: *International Safety Guide for Oil Tankers and Terminals (ISGOTT)*;
— IAPH/ICS/OCIMF: *International safety guide for chemical tankers and terminals*;
— ICS/OCIMF: *Safety guide for terminals handling ships carrying liquefied gases in bulk*;
— SIGTTO: *Liquefied gas handling principles on ships and in terminals*.

8.4. Operational precautions

8.4.1. General requirements

1. Adequate controls over the entry, presence and consequential handling of all types of dangerous goods should be in place for the safety of portworkers.

2. The authority responsible for the port area should be legally able to devise, apply and enforce appropriate rules and standards. International guidance can be found in the IMO's *Recommendations on the Safe Transport of Dangerous Cargoes and Related Activities in Port Areas*.

3. The regulatory authorities, port management, berth operators, shipping companies and portworkers all have various responsibilities. In addition, the many organizations that may be involved with dangerous cargoes even before they reach the port area and ship have a crucial role to play in the ultimate safe handling and transportation. This group includes shippers, packers, those concerned with documentation, consolidators and forwarding agents, collectively called "cargo interests", and all relevant provisions concerning the safety of dangerous goods in the port area should apply to them.

8.4.2. Training

1. All persons involved with the transport of dangerous goods should be appropriately trained. Such persons include persons employed by regulatory authorities, shipping companies and cargo interests, as well as port management, supervisors and portworkers.

2. All persons should receive training appropriate to their duties. Different training will be appropriate for different groups of portworkers.

3. Not everybody should attend the same course. Training should be tailored according to the responsibilities and involvement of the persons concerned. For example, very few persons in the port will need to know the entire *IMDG Code*, although everyone needs to know the part or parts that apply specifically to his or her work, and the relevant port and company rules or instructions and safe systems of work.

4. Specific training on the action to be taken in an emergency involving dangerous goods should be given in addition to the appropriate general awareness and familiarization, and function-specific training.

5. The general training should be designed to provide familiarity with the general hazards of the dangerous cargoes handled and the relevant legal requirements. This should include a description of the classes of dangerous goods and their marking, labelling, placarding, packing and segregation, documentation and emergency response procedures.

6. All portworkers should receive training and instructions on the action they should take in the event of a spillage or other release of dangerous goods.

7. The training should be ongoing and periodically supplemented with retraining, as necessary.

8. All training should be recorded.

9. Relevant training materials include:

— Training unit P.3.1 (Handling dangerous cargoes in port) of the ILO's Portworker Development Programme;

— IMO model course 1.10, Dangerous, hazardous and harmful cargoes;

— *IMDG Code*, Chapter 1.3, recommendations on "Training of shore-side personnel".

8.4.3. Control of entry and presence

1. The entry of dangerous goods into port areas by any mode of transport should be controlled.

2. The port authority should determine any restrictions that may be necessary on:

— classes or quantities of dangerous goods that may be brought into or be present in the port area;

— conditions under which dangerous goods may be present or handled.

3. The port authority should be empowered to prohibit the entry of dangerous cargoes for keeping[1] or storage within,

---

[1] "Keeping" refers to cargoes that are present in a port area after being taken off one means of transport and awaiting the next means of transport to take them on their consigned journey. "Storage" refers to cargoes that are held in the port area awaiting further instructions concerning their sale and/or onward delivery.

or transit[2] through, the port area if it is considered that their presence would endanger life or property because of their condition, the condition of their containment, the condition of their mode of transport or the conditions in the port area.

4. The port authority should also be able to remove or require the removal of any dangerous goods or any ship, package, freight container, tank container, portable tank, vehicle or other cargo transport unit containing such goods within the port area that constitute an unacceptable hazard by reason of their condition or that of their containment.

8.4.4. Notification of dangerous goods

1. The port authority should be notified before dangerous goods are brought into the port area.

2. Unstable substances should not be accepted unless all necessary conditions to ensure safety have been specified and met.

3. The notice required should generally be not less than 24 hours. Special arrangements may need to be made for short voyages and for certain modes of transport, categories and/or quantities of dangerous goods.

4. The notification should include the information specified in Annex 1 of the IMO's *Recommendations on the Safe Transport of Dangerous Cargoes and Related Activities in Port Areas*.

---

[2] "Transit" refers to goods that are destined for another port and are due to remain on board the ship while in the port area.

5. Notification of dangerous goods carried by inward-bound ships should be made by the master, the shipowner or his or her agent. Notification of such goods carried by land-based means of transport should be made in accordance with national legal requirements.

6. The method of notification and the authority to which it should be sent should be determined by the port authority.

7. The port authority should make arrangements for the receipt, checking and assessment of notifications.

8. The port authority should be notified of the dangerous goods on board a ship before its departure. Generally not less than three hours' notice should be given.

8.4.5. Checking the information

1. When notifications of incoming dangerous goods are received, it is important to check that:
— the goods can be handled safely while they are in the port area;
— they are correctly identified and declared;
— they will be kept at an appropriate location;
— any special arrangements, including emergency arrangements, are made.

2. The terminal operator should ensure that:
— packaged dangerous goods entering the terminal have been declared in accordance with national legal requirements as properly identified, packaged, marked, labelled or placarded in accordance with the *IMDG Code*;
— the information supplied by the ship and by cargo interests is verified and, as far as can be ascertained, correct.

3. Packaged dangerous goods entering from the shore side should be physically checked at the port or terminal entrance or some other area designated for the purpose to ensure that the correct labels, placards and other external attachments required by the *IMDG Code* are all present and correct.

4. The checks should be carried out continuously throughout the operational hours of the port. The procedures should include the action to be taken if the documentation or labels/placards, etc., are found to be incorrect. Dangerous goods should not be allowed to go further along the transportation chain until any problems have been corrected or clarified.

8.4.6. Handling and stowage

1. The terminal operator should ensure that dangerous goods are stowed safely, taking into account any segregation required by incompatible cargoes. The general segregation rules contained in the *IMDG Code* should be used for this purpose. However, any other suitable standard may be used, provided that it is effective and that all those concerned with its operation are aware of its provisions.

2. Dangerous goods may be kept in specified areas in sheds, warehouses or on the terminal, or with other cargo. Any of these options may be acceptable, but in each case proper segregation rules should be applied

3. In remote areas, less stringent requirements may be acceptable, but in areas sited near housing, chemical plants or tank farms, more stringent stowage and segregation requirements may be necessary.

4. Irrespective of any other requirements, special care should be taken when handling dangerous goods, whether manually, by lifting appliance or by internal movement vehicle.

5. Class 1 explosives, other than those in Division 1.4S, and Class 6.2 (Infectious Substances) dangerous goods (in the United Nations system of classification), should normally only be permitted to enter the port area for direct shipment or delivery.

6. Special procedures should be drawn up for the transport and handling of explosives. These should take into account the hazards involved, the number of people in the vicinity and the precautions set out in the *IMDG Code*.

7. The terminal operator should ensure that areas where packages of dangerous goods are kept are properly supervised and that such goods are regularly inspected for signs of leakage or damage. Leaking packages should only be handled under the supervision of a responsible person.

8. Nobody should be permitted to open or otherwise interfere with any freight container, tank container, portable tank or vehicle containing dangerous goods without due authorization.

8.4.7. Emergency arrangements in the port area

1. Appropriate arrangements should be made to deal with any emergency that may take place in the port area. At all times these should include:
— provision of appropriate means to raise the alarm both on shore and on board a ship;
— procedures for notifying the emergency services;

— procedures for the action to be taken by all persons;

— provision and availability of appropriate emergency equipment and emergency response information;

— means to determine the identity, quantity and location of all dangerous goods in the port area; this should include their correct technical names, UN Numbers and classifications; these should be made available to the emergency services when required.

2. The relevant emergency arrangements should be drawn to the attention of all persons in the port area.

3. For further guidance, see Chapter 11 and the IMO's *Recommendations on the Safe Transport of Dangerous Cargoes and Related Activities in Port Areas.*

8.4.8. Special provisions

1. The port authority should appoint at least one responsible person who has adequate knowledge of the current national and international legal requirements concerning the safe transport and handling of dangerous goods. That person should have all relevant national and international legal requirements, guidelines, recommendations and other documents concerning the transport of dangerous goods, ships carrying such goods and installations handling, transporting, producing or otherwise using such goods. These documents should be readily available in the port area for reference, and should be kept up to date.

2. Areas where dangerous goods may be present should be designated as areas where smoking and other sources of ignition are prohibited. Electrical equipment in such areas should be explosion-protected, where necessary. The carrying

out of hot work or any other activity that may lead to a fire or explosion hazard in such areas should be strictly controlled.

3. Records should be maintained of the dangerous goods that are present in the port area for use in an emergency.

4. Special areas for the holding and repacking of damaged dangerous goods, or wastes contaminated with dangerous goods, should be provided wherever necessary. All such packages, unit loads or cargo transport units should be immediately moved safely to a special area, and not removed from that area until they have been made safe.

5. The master should ensure that any cargo-handling operations will not hazard any bunkering operation and that these conditions are fulfilled during the entire time the bunkering takes place. The IMO's *Recommendations on the Safe Transport of Dangerous Cargoes and Related Activities in Port Areas* includes a checklist which should be followed before the bunkering operation commences.

8.4.9. Repair and maintenance work

The port authority should require notification before any repair or maintenance work is carried out, either on board a ship or on the shore, which could constitute a hazard because of the presence of dangerous goods. Such work should only be authorized after it has been established that the work can be carried out without creating such a hazard. A "permit to work" system should be used where appropriate.

9. Health

9.1. Health hazards

9.1.1. General requirements

1. Portworkers should be fit for the work which they are employed to carry out. They should be protected from health hazards that may arise from the activity itself, the means to carry out that activity, the work environment or the organization of the work. This part of the code gives examples to assist in identifying the risks and detailing the action that should be taken to avoid them.

2. The health and fitness for work of portworkers who regularly work in areas or on operations known to include health hazards should be regularly monitored by persons competent to do so (see section 9.2). Those carrying out the monitoring process should regularly liaise with those responsible for areas or operations to ensure that the precautions and arrangements for eradicating, reducing or controlling the hazards are effective.

3. Health hazards should be identified, the risks known and evaluated, the dangers to health understood and effective preventive measures put in place to ensure the health of the portworkers concerned. There should be a management system for identifying such risks and a strategy for responding to them. Arrangements for the participation of workers should include health matters.

4. The principal health hazards that can arise from port operations are noise, fatigue, fumes, vibration and exposure to hazardous substances, including cargoes. These hazards

should be controlled in accordance with national legal requirements.

5. Exposure of portworkers to particular hazardous substances should be kept below the relevant maximum 15-minute and eight-hour time-weighted occupational exposure levels for the substances concerned.

6. Portworkers exposed to hazardous materials should be trained, and provided with material safety data sheets. The materials should be adequately labelled with the contents. Workers should be advised as to the precautions to be taken when exposed to these materials.

9.1.2. Dangerous goods and fumigation

1. Health hazards may arise from specialized activities associated with dangerous goods.

2. Great care should be exercised when it is necessary to inspect or sample such goods. Particular attention should be paid to the hazards of the cargo as indicated by the labels or placards and documentation.

3. Cargo transport units that have been transported under fumigation should be declared and should bear the fumigation sign. They should be ventilated before entry into them is permitted. In order to ensure that the atmosphere is safe for entry, it will normally be necessary to test it first.

4. If the cargo, packaging or dunnage in a cargo transport unit is of a category that might need to be fumigated, fumigant residues may still be present in the unit. Precautions before entry should be taken, even though the cargo may not be "dangerous goods" and may not have been declared as being transported under fumigation.

5. When cargo is required to be fumigated within the port area before onward transportation, the operation should be carried out by competent specialists in an area away from normal operations. Precautions should be taken to ensure that the fumigant is confined to the immediate area where it is being applied.

6. Bulk cargoes may also be fumigated:

— in the case of exports:
  - before reaching the port area;
  - when in storage in the port area before loading;
  - when in the ship's hold before the ship sails.

— in the case of imports:
  - before or during the voyage and while still under fumigation;
  - in the port area before onward transportation.

7. Bulk cargoes such as grain which have been fumigated before entry into the port area from shore or from sea should be declared and the port authority should require such a declaration to be made before entry. In addition:

— adequate and suitable measures should be adopted to safeguard the health of portworkers engaged in handling such cargoes;

— such measures should take account of the possibility that fumigant is still present in the cargo.

9.1.3. Dusty cargoes

1. Exposure of portworkers to dust should be prevented as far as is practicable. This should include nuisance dusts for which no specific occupational exposure level has

been assigned. National legal requirements should specify maximum occupational exposure levels of individual dusts and nuisance dusts.

2. Ideally, loading or unloading of dusty cargoes should be totally enclosed. Where this is not practicable, dust emissions should be prevented as far as possible and controlled.

3. Measures to control dust emissions include:

— appropriate design of grabs, hoppers, conveyors and other material-handling equipment;

— enclosure of transfer and discharge points;

— enclosure of operators' cabs;

— local exhaust ventilation;

— suppression by covering or damping.

4. Other measures to limit exposure to dusts should include:

— avoiding the need for portworkers to enter or work in dusty areas;

— restricting the time spent in such areas;

— provision of appropriate respiratory protective equipment, such as helmets providing a continuous supply of clean filtered air;

— ensuring that respiratory protective equipment is worn when necessary.

5. The air supply to an enclosed cab or control room should be from a clean source and filtered as necessary.

6. Accumulations and residues of dust should be regularly cleaned up by an appropriate method.

7. Some dusts, such as grain, can have a sensitizing effect that can induce changes in the respiratory system such as asthma or other medical conditions. Portworkers who may be sensitized should not work in areas where they may be exposed to such dusts.

8. Other cargoes can also give off dust that may be harmful in enclosed spaces. These can include some forest products and scrap metal.

9. Exposure to asbestos fibres can give rise to cancer and mesothelioma, and should be prevented. All asbestos cargoes should be properly packaged.

9.1.4. Other cargoes

1. Some cargoes, including mouldy cargoes, may present risks of infection of portworkers. Portworkers handling such cargoes should be under appropriate medical supervision and be provided with, and use, relevant personal protective equipment.

2. Exposure to hides, skins, fleeces, wool, hair, bones or other parts of animals can give rise to anthrax or other animal-related diseases which may be transmitted to and be harmful to humans. Such cargoes should be disinfected and certificated by a competent authority before shipping in accordance with national legal requirements. When the risk of anthrax may be suspected, special precautions, including the use of personal protection equipment and medical supervision, should be taken.

3. Some cargoes may bring with them insects, snakes and other creatures, and portworkers should be alert to the

dangers of being bitten. In the event of such a bite, they should receive medical treatment immediately.

4. Radioactive materials should be contained by packaging appropriate to the risk. Correctly packaged cargoes of this type are safe for transport workers, provided that total quantities do not exceed international transport index limits.

9.1.5. Noise

1. Noise can be emitted from engines and transmission equipment fitted to lifting appliances and vehicles, and can be heightened when the equipment is being used in a shed, warehouse or ship's hold. Noise levels may affect the equipment operator and/or portworkers that work with or in the vicinity of such equipment when it is being used.

2. In coordination with the workplace safety committee, noise levels should be periodically monitored and sources of excessive noise identified.

3. Noise levels should be controlled at source whenever it is practicable to do so.

4. Noise levels, as defined by national legal requirements, should be specified when new equipment is ordered. The noise specification should be as low as possible.

5. The need to work in noisy areas should be avoided or minimized as far as possible. Appropriate hearing protection should be supplied and worn when necessary.

6. When appropriate, consideration should be given to the periodic monitoring of hearing loss of portworkers.

9.1.6. Fatigue

1. Fatigue can affect health, safety and work performance. Regular breaks should be incorporated into work periods. Excessively long shifts or work periods should be avoided.

2. If it is necessary to work an abnormally long shift, it is essential that an adequate period of rest be provided before the start of the next period of work, particularly overnight.

9.1.7. Fumes

1. Exhaust fumes emitted by terminal machinery, trade cars, passenger vehicles, ro-ro vehicles and trade wheeled cargo vehicle exhausts can present health risks to portworkers from:

— carbon monoxide (the main component of fumes from petrol engines);
— carbon dioxide (the main component of fumes from diesel engines);
— polycyclic aromatic hydrocarbons (PAH);
— oxides of nitrogen;
— sulphur oxides;
— aldehydes;
— particulate matter, e.g. soot.

2. The composition of exhaust fumes and the risks from them vary with the type of engine, the fuel being used, and the age and level of maintenance of the engine. The volume of exhaust fumes in the atmosphere will depend on the number of engines running at any one time and the level and efficiency of ventilation.

3. Hazardous levels of fumes can affect those in the immediate vicinity, especially if the area is enclosed or if the fumes are concentrated at one particular point.

4. Preventive measures include:

— regular scheduled maintenance of all terminal vehicles, including engine tuning and exhaust systems;

— ventilating places where vehicles operate by natural or mechanical ventilation to ensure safe levels;

— switching engines off when vehicles are standing for long periods;

— making initial fume assessments of individual terminal vehicles and shore-side premises where vehicle fumes may present a problem;

— preparing fume profiles of each hold of a ship in which vehicles may be operated on a regular basis;

— preparing a plan to ensure that fumes from such vehicles on premises and in holds do not exceed safe levels, and specifying the maximum number of engines allowed to run at any one time;

— using electric, LPG- or LNG-driven vehicles where appropriate.

5. Profiling of holds can typically consist of taking readings of fume levels in each hold at hourly intervals throughout the loading/unloading process. It is not normally necessary to take readings for each component of the vehicle exhaust fumes; only carbon monoxide and carbon dioxide readings need be taken. If it can be shown that the levels of those components are safe, it can normally be assumed that the other components are equally safe. If there is any doubt, an industrial hygienist or other expert should be consulted.

6. Profiling of holds should be carried out with all available ship's ventilation in operation and with the maximum number of vehicle engines consistent with operational procedures running at any one time.

7. Each set of readings should consist of at least six measurements, two at each end and two in the middle of the hold taken at about 1 m above the deck. Alternatively, individual monitors may be used.

8. The levels measured should not at any time exceed the eight-hour time-weighted occupational exposure limits for carbon monoxide and carbon dioxide. If the measured levels are reduced to acceptable levels, the levels of the other components, which are present in very small amounts, should also be at safe levels. In such circumstances, most portworkers may enter and leave the hold without being required to record their exposure times or wear respiratory protective equipment.

9. If higher levels are measured in a hold, the shore-side employer and the ship's officers should collaborate to ensure that they are reduced to an acceptable level. This may be achieved by increasing ventilation, adding portable ventilators or reducing the number of engines running at any one time.

10. Once a ship has been profiled, and it can be shown that all holds or areas where portworkers may work are within safe limits at all times while the ship is loaded or unloaded in the normal way, further profiling should not be necessary unless the purpose or configuration of the ship changes.

11. This procedure should be satisfactory for most portworkers working on ships as they move in and out of the hold

during cargo operations. However, certain portworkers, such as those engaged on lashing or unlashing vehicles on ro-ro ships or forklift truck operators in sto-ro operations, may be present and vehicle exhausts emitted for the entire work period. There may also be others who work for extended work periods of over eight hours. As occupational exposure levels are based on time exposure, more detailed calculations will need to be made in these circumstances to safeguard the health of such workers, and personal protection may be necessary.

12. During operations in holds, it is essential to ensure that:

— all available ship's ventilation is in operation;

— the ventilation functions correctly with exhaust fans not reversed, and air ducts are not covered or obstructed at either end;

— doors, ramps and other openings in the hull are open to permit natural ventilation;

— only the stipulated number of engines are being run at any one time.

13. Appropriate precautions should be taken to protect the operators of quayside cranes from fumes from ships' funnels.

9.1.8. Vibration

1. Hand/arm and whole-body vibration generated by powered hand tools and machinery can affect human health. Vibration levels should be measured and kept below nationally recognized maximum and eight-hour exposure levels. Operators of terminal plant are most likely to be adversely affected by whole-body vibration levels.

2. An assessment of the risks from vibration should be based on readings of each piece of terminal operating equipment and should lead to the preparation of a plan that will ensure that such equipment remains safe for its operators.

3. It should be ensured that all new equipment is designed to minimize vibration to below nationally recognized levels.

9.1.9. Abnormal environments

Where portworkers are engaged in abnormal environments such as extremes of temperature, or where the wearing of respiratory equipment is essential, they should be relieved at suitable intervals for rest in fresh air.

9.1.10. Other health aspects

1. Where portworkers are accidentally exposed to health hazards, their health should be checked by persons competent to do so.

2. Where portworkers handle harmful substances, they should change their outer clothes and thoroughly wash their hands and face with soap or some other suitable cleaning agent before taking any food or drink.

3. Health protection surveillance should be considered for special groups, e.g. juveniles, older portworkers, female portworkers, disabled persons and insulin-dependent workers.

4. Special attention should be paid to risks from manual handling, especially heavy loads. Portworkers should not be engaged on such activities without suitable medical assessment and training in the skills necessary to carry out manual handling safely.

9.1.11. Ergonomics

1. Workplaces, work systems and work equipment should be designed, constructed and maintained in accordance with good ergonomic principles. When necessary, specialist advice should be obtained.

2. Bad design of the operator's cab and poor posture can affect the health of portworkers, especially if they are spending most of their working time in the same position. This includes:

— the layout and positioning of the cab;
— the climate inside the cab;
— daylight and illumination;
— possible obstruction of view from within the cab;
— noise;
— speech intelligibility;
— positioning of displays and other communication means;
— positioning of the hand and foot controls, and the design and positioning of the operator's seat.

These should all be taken into account when designing and arranging operators' cabs.

3. When appropriate, the continuous time spent on a task should be limited, possibly by job rotation or other suitable relief.

9.2. Occupational health services

9.2.1. General principles

1. The recruitment of portworkers should be in accordance with their tasks in the port area.

2. Good prevention practices include the ability to detect and monitor work-related injuries or occupational diseases. This may be by means of instituting a medical evaluation programme, which is highly recommended. The benefits of such a programme will be to provide appropriate protection to workers in the workplace.

3. In accordance with national legal requirements, the medical evaluation programme could include elements pertaining to:

— respect for patient/doctor confidentiality;

— occupational hazards;

— adequate follow-up.

4. The development and implementation of a medical evaluation programme should be carried out in consultation with the employers and workers and their representatives.

5. Occupational health services should establish and maintain a safe and healthy environment to facilitate optimal physical and mental health in relation to work, and should also provide advice on adapting work to the capabilities of workers in the light of their state of physical and mental health. These services, which may be supplemented by others listed in Article 5 of the Occupational Health Services Convention, 1985 (No. 161), are the following:

— provision of first-aid and emergency treatment;

— treatment and care of urgent cases;

— surveillance of workplaces and conditions from the standpoint of the health and fatigue of the workers;

— periodic training of first-aid personnel;

— promotion of health education among portworkers;

— cooperation with the competent authority in the detection, measurement and evaluation of chemical, physical or biological factors suspected of being harmful.

9.2.2. First-aid personnel

1. First-aid stations should be provided where less serious injuries can be treated and from which injured portworkers can if necessary be conveyed rapidly to a centre where they can receive more comprehensive medical attention.

2. First-aid stations should be clearly marked and should contain first-aid equipment in accordance with national legal requirements.

3. In general:

— except in emergencies, first aid in case of accidents or sudden illness should be given only by a medical doctor, a nurse, or a person trained in first aid and possessing a first-aid certificate acceptable to the competent authority;

— adequate means and personnel for giving first-aid treatment should be readily available during working hours at places where port work is carried out;

— severely injured portworkers should not be moved before the arrival of a doctor or other qualified person, except for the purpose of removing them from a dangerous situation;

— all injuries, however slight, should be reported as soon as possible to the nearest first-aid person or room.

4. Where portworkers may be exposed to the risk of injury from corrosive substances:

— suitable first-aid facilities such as eyewash bottles and means of drenching with water should be provided and kept readily available;

— notices giving suitable first-aid advice should be displayed.

5. Employers in the ports may organize first-aid facilities in collaboration with each other. For first aid to be effective, close coordination between all organizations concerned is essential.

9.2.3. Personnel providing occupational health services

1. Personnel providing occupational health services should enjoy full professional independence from employers, workers and their representatives, where they exist, in relation to the functions listed in section 9.2.1, paragraph 5.

2. The competent authority should determine the qualifications required for the personnel providing occupational health services, in the light of the nature of the duties to be performed and in accordance with national law and practice.

10. Personnel welfare facilities

10.1. General provisions

1. Adequate personnel welfare facilities should be provided and be available at all times to portworkers at or near the area in which they work.

2. Toilet facilities, washing facilities, cloakrooms, mess rooms, canteens, hiring halls, waiting rooms and any other personnel welfare facilities should be:
— suitably located and of an appropriate size and construction;
— fully enclosed, if on shore;
— provided with floors, walls and ceilings that are easy to clean;
— well ventilated and lighted, and, if necessary, heated or air-conditioned;
— equipped appropriately for their purpose;
— in the charge of a responsible person;
— maintained in a clean, sanitary and orderly condition;
— protected against rats and other vermin.

10.2. Toilet facilities

1. Suitable and sufficient toilet facilities should be provided for the use of all portworkers.

2. All toilet facilities should comply with national health and hygiene requirements and be fitted out in accordance with local custom.

3. Toilet facilities should be located at regular intervals, as far as is practicable, throughout the port area. The

facilities may be located near to sheds or other buildings so as not to cause obstructions on quay areas.

4. At least one toilet should be available for portworkers on board ship, where practicable.

5. Toilets and urinals should be of the water-flush type, wherever possible.

6. Floating cranes, grain elevators, bunker machines and similar installations on which or by means of which port work is carried out should be provided with at least one water closet.

7. The number of toilets provided should be based on the maximum number of persons expected to work in an area. As a general rule, a set of toilet facilities should be provided for each berth or at least every two berths. Each set should comprise a water closet for every 25 or 30 workers. This may mean two water closets per berth, or four if the set is shared by two berths. Each water closet should be supplemented by an adequate number of urinals.

8. Separate toilet facilities for each sex should be provided unless the toilet facilities can only be occupied by one person at a time.

9. All toilet facilities should be properly enclosed and easy to clean. A floor drain with a water seal should be provided in each toilet to facilitate flushing the floor.

10. Each water closet on shore should be under cover and occupy a separate compartment installed in a special toilet room. Each compartment should be provided with a separate door fitted with a latch on the inside.

11. Urinals should be of suitable width and preferably consist of a row of stalls. If the urinals are of a smaller type (cuvettes) they should be adequately separated by side partitions.

12. For personal cleansing, an adequate supply of toilet paper or, where local custom requires, water should be provided.

13. Adequate washing facilities, including soap and means of drying hands, should be provided in or adjacent to each toilet area.

14. Consideration should be given to the need to provide toilets equipped for use by disabled persons.

10.3. Washing facilities

1. Suitable and sufficient washing facilities should be provided for all portworkers.

2. There should be at least one washing facility for every ten portworkers who are likely to use them at the same time.

3. If portworkers of both sexes are employed, separate washing facilities should be provided for each sex.

4. Each wash place should have:

— a sufficient flow of clean hot and cold or warm water;

— an adequate means for removing waste water;

— a sufficient supply of suitable non-irritating soap or other cleanser;

— suitable means for drying; the common use of towels should be prohibited.

5. Where portworkers are exposed to skin contamination by toxic, infectious or irritating substances, oil, grease or dust, at least one shower should be provided for every six regularly employed workers who are exposed to such contamination and cease work at the same time. Each shower should have a supply of hot and cold or warm water.

6. Showers should be enclosed in individual compartments, with the entrance suitably screened.

7. Hooks or other facilities for clothing and towels should be provided for persons taking showers.

8. Shower equipment should be thoroughly cleaned at least once a day. An effective disinfectant should be used to destroy fungi.

9. Washing facilities should not be used for any other purpose.

10.4. Clothing accommodation

1. Suitable and sufficient cloakrooms should be provided for all portworkers.

2. Cloakrooms should be provided with:
— individual lockers, preferably of metal, with adequate ventilation for the storage of clothes;
— separate storage facilities for workers' working and street clothes;
— suitable facilities for changing;
— suitable facilities for drying wet clothes;
— benches or other suitable seating arrangements.

3. If portworkers of both sexes are employed, separate cloakrooms should be provided for each sex.

4. When women are employed and no rest room is available, some suitable space in the men's cloakroom should be provided. This space should be properly screened and suitably furnished.

5. Cloakrooms should not be used for any other purpose.

6. Suitable arrangements should be made for disinfecting cloakrooms and lockers in accordance with the requirements of the competent health authority.

10.5. Drinking water

1. An adequate supply of cool and wholesome drinking water should be provided and be readily accessible to all portworkers. All water supplied for drinking purposes should be from a source approved by the competent health authority and controlled in the manner prescribed by that authority. If a supply of wholesome drinking water is not available, the competent health authority should give the necessary directions for making available water safe for human consumption.

2. An adequate number of drinking water outlets should be provided. These should be protected from damage and dirt.

3. Drinking water outlets should be clearly identified as such by a suitable notice stating "Drinking water". The notice should conform to national legal requirements.

4. No confusion with outlets of water that is not suitable for drinking should be possible. If there is any scope for such confusion, outlets of water that is not fit for drinking should be clearly identified by a notice stating that the water

is not fit to drink. Where appropriate, pictorial signs should be used.

5. Where practical, hygienic drinking fountains should be provided.

6. The use of common drinking cups should be prohibited.

7. In places where it is not possible to provide a piped supply of drinking water, such as on tugs, lighters or other harbour craft, drinking water should be provided in sealed bottles or in suitable closed containers clearly marked "Drinking water". The containers should be properly maintained and replenished as necessary. Drinking water should not be contained in barrels, pails, tanks or other containers from which the water has to be dipped, whether they are fitted with covers or not.

10.6. Mess rooms and canteens

1. If portworkers regularly take part in port work on shore or on a ship moored to the shore, suitable mess rooms or canteens on the shore for the consumption of food and beverages should be provided for portworkers to take their breaks and eat.

2. Floors of mess rooms and canteens should be constructed of, or covered with, material that is impervious to water and easily washable.

3. Mess rooms and canteens should be provided with:
— tables with impervious surfaces that are easy to clean;
— suitable chairs or other seating facilities with back rests;
— separation of smokers from non-smokers;

— facilities for heating food and boiling water;
— a supply of clean drinking water;
— covered receptacles for the disposal of waste food and litter. Receptacles should be emptied after each meal and thoroughly cleaned and disinfected;
— adequate facilities for cleaning utensils, tables, chairs, etc.;
— facilities for hanging wet-weather clothing or other outer clothing during breaks.

4. Adequate toilet and washing facilities, including soap and means of drying hands, should be provided in or adjacent to each mess room and canteen.

5. Mess rooms and canteens should be kept in a clean and orderly condition.

6. Mess rooms should not be used as workrooms or storage rooms.

7. The sale or consumption of alcoholic beverages should not be permitted in canteens or mess rooms.

8. The consumption of food or beverages in areas where hazardous materials are being handled or kept should be prohibited.

10.7. Hiring halls and waiting rooms

1. In ports in which the hiring of portworkers takes place on a daily basis or at other frequent time intervals, appropriate hiring places should be provided where employers and workers can meet to agree terms and make up gangs.

2. In such ports, suitable hiring halls or call stands should be provided for the accommodation of portworkers

while they are waiting to be allocated port work. The equipment of these hiring halls or stands is often prescribed in national or local legal requirements relating to the hiring of workers.

3. The hiring place should usually consist of a large hall in which workers gather and employers make offers of employment. The hall should include desks where workers can register. The labour inspectorate should also have an office in the hiring place so that it can more easily keep watch for irregularities.

4. Hiring halls should include suitable areas or rooms in which workers can wait between calls, or between the end of a call and the start of work. These areas or rooms should include adequate seating accommodation, and toilet and washing facilities.

11. Emergency arrangements

11.1. Emergency arrangements on shore and ship

11.1.1. General requirements

1. Many types of emergencies are possible in port areas, and in many countries the development, publication, exercise and regular review of emergency plans in ports is a legal requirement. General advice is given by the IMO's *Recommendations on the Safe Transport of Dangerous Goods and Related Activities in Port Areas*, the OECD's *Guidance concerning chemical safety in port areas* and the UNEP/IMO programme *Awareness and Preparedness for Emergencies at Local Level (APELL) for Port Areas*.

2. Appropriate training or instruction of portworkers on the action they should take in an emergency is essential.

3. Each type of potential emergency that could occur in port areas should be considered when preparing appropriate emergency arrangements.

4. Emergency arrangements and emergency plans (see section 11.2.5) should cover all foreseeable emergencies, from minor mishaps to major incidents. They should be capable of increasing response appropriately as an incident develops.

11.1.2. Injuries and ill health

1. Arrangements for emergencies should include a suitable number of first-aid boxes and first-aid personnel (see section 9.2.2) and readily available means to transport more serious cases to hospital. Some ports have ambulances staffed by paramedics (persons trained to assist medical

professionals and give emergency medical treatment) based within the port area, while others rely on the local community ambulance service. In each case, it should be very clear how the service is contacted. The emergency telephone number should be easily remembered.

2. First-aiders and ambulance personnel should be capable of safely reaching people who are injured, wherever they may be.

11.1.3. Rescue

1. If workers become ill or are injured in places with difficult access and cannot get themselves back to where they can receive help, it may be necessary to rescue them. Such places may include:

— holds of bulk carriers with access only available by hold ladder;

— tops of lighting towers some 50 m high with access only by vertical ladders;

— dry dock pumping pits 25 m deep with access only by staples on the pit walls;

— cabs of container or dry bulk transporter cranes;

— jibs of general cargo cranes;

— outboard gangways of large container ships beyond the reach of the crane;

— water in the port (see section 11.1.7).

2. In each case, the situation should be assessed and the need for a possible rescue considered. Where necessary, the means of carrying out the rescue should be planned taking into account the need to prevent further injuries during

rescue that could result from lack of oxygen, hazardous substances, electricity or other hazards.

3. The possible need for special equipment should be considered. Once rescuers reach a casualty, special lifting/lowering devices and harnesses are often needed for evacuation. Plans should assume that the casualty is unable to assist in any way. Any special equipment should be light and easily transported. It may have to be carried or lifted up and down vertical ladders, possibly following a complete loss of electrical power. The equipment should be capable of being erected or deployed with a minimum of delay. Exercises in the use of the equipment should be held at regular intervals.

11.1.4. Property damage

1. In many cases of property damage, emergency action may be necessary to prevent potential injuries by making the site safe and recovering equipment before repairs can be undertaken.

2. The emergency arrangements should take account of the possible need for heavy lifting equipment and other specialized plant, and persons with particular or specialist expertise.

11.1.5. Fire

1. Emergency arrangements in the event of fire should be additional to the fire precautions described in section 3.1.4 and the various steps taken to prevent the outbreak of fire, such as fire protection of buildings, control of flammable substances and materials and sources of ignition including smoking, and regular inspection of premises and operations.

2. If a fire is discovered, the alarm should be raised immediately; apparently trivial fires frequently develop into serious fires.

3. The emergency plan should set out the action to be taken when the alarm is raised. This should include alerting relevant emergency services. The action to be taken may well vary between different groups in different locations.

4. When evacuation of an area is necessary, all workers should leave the area immediately by the nearest safe route and go to the appropriate fire assembly point. At the fire assembly point, a check should be carried out to ensure that nobody is missing.

5. Fire extinguishers should only be used by persons who have had appropriate training and experience in their use and when it is safe for them to do so. Persons using fire extinguishers should be aware of circumstances when the use of inappropriate extinguishers or equipment could be dangerous. This includes the use of water on electrical equipment and on materials that react with water.

6. Appropriate emergency access for trained firefighters and their equipment, and means of escape in case of fire, should be kept clear at all times.

7. The dangers to workers in the event of fire demand urgent positive action following the discovery of a fire. Fire drills should be carried out at appropriate regular intervals.

8. Arrangements in the event of fire should include arrangements relating to fires on ships and the action to be taken by ships in the event of fire on shore. These should cover fires on ships anywhere within the area of responsibility of the port authority.

9. Fire precautions and emergency arrangements in the event of fire should be coordinated throughout the port area in consultation with the local fire authority. This may be under the lead of the overall port or port authority in accordance with relevant local by-laws or other legal requirements. When appropriate, specific fire precaution measures should be devised in consultation with relevant bodies and specialists.

10. Where attendance by different fire authorities may be necessary owing to the boundaries between their areas of responsibility, it is essential to ensure that no confusion can arise in the event of an incident on or near the boundary (figure 110). This is particularly likely to occur when such boundaries run along rivers.

Figure 110. Boundaries in a river or estuary

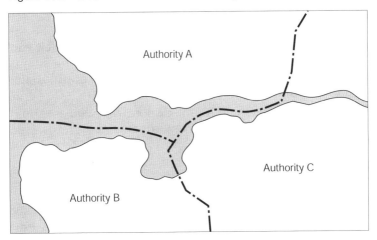

11.1.6. Cargo spillage

1. Spillage of cargo containing dangerous goods may pose a threat to persons in the immediate area. Emergency arrangements should include safe means of identifying the cargo, isolating a leak and, if necessary, rendering it harmless.

2. Hazardous spillages should only be dealt with by trained personnel. Such personnel may be from local emergency services, other specialists or portworkers appropriately trained to deal with low-level emergencies. In every case, the immediate action should be:

— evacuation of the area;

— safe removal of any casualties;

— identification of the spillage.

3. Arrangements to deal with cargo spillages should take into account the fact that it may be necessary to deal with cargo spillages or leakage that occur on board a ship during a voyage when the ship enters the port.

4. Whenever possible, an isolated area should be set aside to which a leaking receptacle, container or vehicle can be taken. Such areas should be bunded with drain sumps connected to sealed systems or interceptors, as appropriate, to prevent contamination of the nearby waters.

5. Any arrangement for the disposal of spillages should take into account potential environmental hazards (see Chapter 12). Sweeping or washing residues over the edge of the quay should be prohibited.

11.1.7. Falls into water

1. By the nature of ports, falls into water are a commonplace hazard, and not all portworkers who may fall into

water may be able to swim. Means by which such persons can rapidly escape from the water or be rescued should be provided.

2. The survival of workers awaiting rescue from water will be aided by the wearing of appropriate buoyancy aids or life jackets, and by the availability of quayside ladders (see section 3.3.5) and life-saving equipment, including chains, handholds or other means to enable persons to support themselves in the water (see section 3.3.6).

3. The emergency arrangements should take into account the fact that it will not be appropriate for many workers to wear buoyancy aids or life jackets at all times. It should be recognized that buoyancy aids only provide support to conscious wearers who are able to swim and help themselves, but life jackets will support their wearers, particularly those who are unable to swim, injured, exhausted or unconscious. Buoyancy aids may be suitable in sheltered water where there are other persons in the vicinity and rescue can be expected very quickly. Such garments are lightweight and offer very little hindrance to movements. Life jackets are the most effective means, and modern designs of the self-inflating type allow them to be worn by workers undertaking manual tasks such as the mooring of ships. Life jackets should generally be worn when working afloat.

4. Speed is essential for the rescue of persons in the water, as it can prevent a fall into the water from having tragic results. Means of rescuing should, therefore, be capable of being deployed very quickly. Delay may result in workers clinging to a fixed floating object after a simple fall being affected by fright, cold water, currents and tide, and may soon make them lose consciousness and let go.

5. Suitable rescue boats should be available as necessary, particularly where there is a fast-flowing current and the work is done on board barges or other small ships away from the quay. Rescue boats should be capable of being crewed by at least one trained first-aider and mobilized rapidly. When the ship being worked is moored at a distance from the quay, the boat for rescue purposes may be the tender used to carry the workers, with at least one suitable person responsible for manning the boat in the event of an alarm. The use of powerful rubber dinghies with very small height above the water makes it possible to grasp victims and haul them on board without difficulty, and as their hull is fairly flexible there is less likelihood of it injuring the casualty if he or she is struck by it. Rescue boats with higher freeboards should preferably have recovery devices and/or stern platforms and ladders.

6. When victims have been taken out of the water, they should be warmed, their wet clothes should be taken off if possible and they should be wrapped in blankets or other suitable wrapping.

7. If a victim no longer seems to be breathing, artificial respiration should be applied by the mouth-to-mouth method or, if that is not possible, by the Holger-Nielsen method. Resuscitation may be stimulated by using a bladder to administer oxygen or by giving injections, but only appropriately qualified persons with special training should give such treatment.

11.1.8. Failure of services

Consideration should be given to the effect of a failure of essential services, such as electricity or communications, that could affect a limited area or the entire port premises. The failure may be part of a wider emergency, such as a severe storm, or an isolated event, such as the severing of cables during construction operations.

11.1.9. Severe weather and other natural hazards

1. Ports may suffer from a variety of natural events. These include:

— high winds and severe storms;
— flooding from tides, river water, land water or a combination of both;
— snow and ice;
— temperature extremes;
— earthquakes;
— volcanic eruptions.

2. Arrangements should be made with a reliable weather service to obtain warnings of adverse weather conditions in good time to enable appropriate action to be taken before the arrival of the adverse weather. The action may include:

— stopping cargo handling;
— moving and securing large cranes and other objects likely to be affected by the predicted conditions;
— deploying sandbags and other flood-protection equipment;
— evacuation.

3. Snow and ice are likely to result in slippery surfaces for people and machines, and a film or covering of ice may develop on some cargoes, making them heavier, very slippery

to walk on and difficult to handle. Particular care should be taken in such conditions, and suitable thermal clothing and good footwear with slip-resistant soles should be provided and worn. Other precautions may include the availability of stocks of rock salt to treat roads, pathways and cargo-handling areas, regular mechanical or manual sweeping of outside surfaces, and chains or studs on tyres.

4. Some ports regularly operate in temperatures below -40°C and over +40°C. Exposure to extremely high or low temperature is likely to affect the ability of portworkers to continue to work safely and without endangering their health. Appropriate precautions should be taken, particularly if such conditions are only experienced occasionally. Precautions may include limiting the time workers are outside in any one period, and arranging a readily available supply of clean drinking water and suitable clothing.

5. The benefits of pre-planning should be clear, with foresight rather than hindsight.

11.1.10. Major hazard installations

Some ports may be major hazard installations because of the storage or other activities of specified threshold quantities of hazardous substances in the port, or be adjacent to such an installation. In such cases the relevant national legal requirements and guidance given in the ILO code of practice *Prevention of major industrial accidents* should be followed.

11.2. Emergency planning

11.2.1. General requirements

1. Emergencies should be anticipated, and arrangements for them prepared and published as an emergency

plan. Guidance on emergency planning at ports is contained in the IMO's *Recommendations on the Safe Transport of Dangerous Goods and Related Activities in Port Areas*, the UNEP/IMO programme *Awareness and Preparedness for Emergencies at Local Level (APELL) for Port Areas* and the OECD's *Guidance concerning chemical safety in port areas*.

2. The port should have an overall emergency plan. In addition, each enterprise in the port should have its own emergency plan. All such plans should be compatible and harmonized with one another, and should include arrangements for alerting the port authority to emergencies in individual terminals. References to port emergency plans (section 11.2.5) below are equally applicable to ports and terminals.

3. Port emergency plans should be harmonized with national emergency plans, local community emergency plans and the plans of other enterprises such as local industrial plants or in-port-area airports.

11.2.2. Scope

1. The overall objectives of an emergency plan are to:
— contain and control emergency incidents;
— safeguard people in the port and neighbouring area;
— reduce the effects of an incident, and minimize damage to property and the environment.

2. The plan should cover the entire port area and all activities within that area. As such, it should include, as appropriate, the harbour area within the port's limits. On the shore side, it should cover the quayside, the ship alongside, terminal areas, roadways, lock sides (where relevant), administrative buildings and other premises within the port area.

3. Some ports have operational heliports or airports situated within or close to the port area. In such situations, the possibility of air emergencies should also be considered.

4. Port security access arrangements should be harmonized with emergency response services to ensure that there is no delay of access due to port facility entrance closures or controls.

11.2.3. Responsibilities

1. Prime responsibility for the emergency plan should rest with the port authority. The authority should develop and publish it in consultation with other interested organizations.

2. Within the port authority, it should be clear where the responsibility lies for developing and updating the plan. This is often with the harbour master, but may lie elsewhere.

3. Within a terminal, the owner or operator is responsible for the production of the emergency plan, but this duty usually devolves upon the terminal manager.

11.2.4. Liaison

1. There should be full and proper consultation with all interested parties during the preparation of emergency plans, including those that have their own emergency plans. Port emergency plans should take into account the possibility of an event in the port affecting neighbouring premises and an event in neighbouring premises affecting the port premises.

2. The organizations and groups to be consulted in developing the plan and in any revision of it should include, as appropriate:

— berth operators;
— port users;
— government departments;
— local communities;
— workers;
— waste disposal authorities;
— water authorities.

11.2.5. Emergency plans

11.2.5.1. General requirements

1. The plan should be concerned with four factors, namely:

— the hazard and nature of an event and its possible extent;
— the risk and probability of its occurrence;
— the consequences and the possible effect on people and the environment;
— the means and actions to be taken to minimize the consequences of the event.

2. An effective port emergency plan is one that clearly sets out the actions to be taken in simple terms. It should be flexible, and capable of responding effectively to any emergency that might arise. The framework should include:

— identifying responsible persons to take control;
— provision of a control centre;
— arrangements to assess the situation;
— the initiation of remedial action;
— provision for monitoring events as they develop.

3. The plan should be based on the particular circumstances of the port, including its geographical position, its cargoes, the numbers of people employed in the port, the possible presence of the public as passengers or for other reasons, and the possible proximity of schools, hospitals and housing outside the port boundary.

4. The basic plan should be concerned with overall procedure and control. It should be supplemented by more detailed plans for specific types of emergencies. Although each should follow the same overall procedures and control, the detailed planned actions will necessarily differ; for example, response to a severe high wind will be very different from the response to a major fire and explosion.

5. The plan should cover all types of emergencies that could occur in the port and include responses that are appropriate to the severity of the incident. The response should be capable of being scaled up or down as an incident progresses.

6. Simple routine responses will be appropriate for many minor emergencies. For major emergencies, a major emergency plan will be necessary.

11.2.5.2. Essential personnel

1. It is essential that the plan identify appropriate key personnel to control an emergency and assign specific responsibilities to them.

2. The two key persons are the *main controller* and the *incident controller*. The plan should specify who should undertake each role in an emergency.

3. The main controller should have overall responsibility for directing operations. This could be the port captain, harbour master or the chief of the fire services.

4. The incident controller should take charge at the scene of the incident and, in the initial stages, may also act as main controller. Consequently, the incident controller should have a comprehensive knowledge of the emergency plan and the situation within the port at any given time. This could be the operations shift manager.

5. All appointments should be formal and made in writing. The appointment should be specific by name or by position and should not be left to be determined at the time of an incident.

6. Provision should be made for:

— sickness, leave or other absence (need for deputies to be specified);

— incidents occurring at any time of the day and night on any day of the year, irrespective of whether the port is operational or not;

— continuous cover over every work period.

7. Other groups of important personnel include those who will advise the main controller and implement his or her decisions, such as representatives of shipping companies, berth operators, port users, and specialist advisers, as well as those who will act as marshals or runners, or undertake emergency work in other supporting roles. All should be clearly identified and able to be rapidly contacted in an emergency.

11.2.5.3. Roles

1. On being made aware of an incident that is, or could develop into, a major emergency:

— the incident controller should proceed to the scene, and assess the nature and size of the occurrence and whether it constitutes a major emergency or might do so. On deciding that there is or could be a major emergency, the incident controller should activate the major emergency plan, take charge of the area and assume the duties of the main controller until the main controller arrives and takes over;

— the main controller should proceed to the emergency control centre, take over control and declare a formal emergency situation, if and when appropriate, and then take appropriate action depending upon the situation.

2. The main controller may call out key personnel and directly exercise operational control of parts of the port area outside the affected area, continually review and assess developments, direct such closing of berths and their evacuation as may be necessary, liaise with chief officers of the fire and police services and with the local authority, and ensure that statements are issued to the relevant authorities and the news media. A log should be kept of the entire emergency.

3. The incident controller's first steps should be to safely close down and evacuate the immediate area of the incident, and any other areas likely to be affected, and to ensure that the emergency services have been called and key personnel summoned.

4. The incident controller's main responsibility is to direct operations at the scene of the incident, and this will

include rescue and fire-fighting operations (until taken over by the emergency services), searching for casualties and evacuation of all non-essential personnel. The incident controller should set up a communications point with radio, telephone, electronic or messenger links, as necessary, with the emergency control centre, advise and inform the emergency services as required, and brief the main controller on a regular basis.

11.2.5.4. Emergency control centre

1. An emergency control centre should be established from which the main responses to the emergency should be directed and coordinated. The main controller, key personnel and senior officers of the emergency services should be based there. The centre should be:

— located, designed and equipped to remain operational throughout an emergency;

— equipped to receive and send information and directions to and from the incident controller, other parts of the port area and areas outside it;

— equipped with a sufficient number of internal and external telephones, at least one of which should be ex-directory or outgoing only;

— provided with detailed maps of the port area, an up-to-date call-out list of key personnel, the site of major hazard sites within or near the port area, and the location of relevant emergency equipment such as safety equipment, fire-water systems and fire-extinguishing materials, neutralization materials, absorbent materials and oil booms.

2. Details should be retained of external specialists, sources of specialist equipment, advice and information, and publications that might contain valuable information such as the *International Maritime Dangerous Goods Code (IMDG Code)*, its supplements the *Emergency Schedules (EmS)* or the *Medical First Aid Guide (MFAG)*, or a suitable computer database. Up-to-date home and out-of-hours telephone numbers of all key personnel, external specialists and equipment persons should be kept in the centre.

3. Facilities should be provided for the media during an emergency. These should be separate from the emergency control centre to prevent media requests distracting those in operational control of the emergency.

11.2.5.5. Publication, exercise and review

1. The emergency plan should be published within a reasonable time of finalization, and made available to all those who are concerned with its contents and activation in an emergency.

2. Plans should also be exercised. The use of plans in live incidents can be taken as a test of the relevant part of the plan. In the absence of such incidents, it is recommended that the plan, or some part of it, should be exercised on a regular basis. The frequency should be determined in the light of local circumstances, but it is recommended that it should be no longer than three years.

3. All emergency plans should be subject to review. The use of the plan in a live incident can be used as part of a systematic review of the operation of the plan. In the absence of live incidents, the most common review period is 12 months, although this would depend on local circum-

stances. It is recommended that the review period should generally be no longer than three years.

11.3. Emergency equipment

1. Suitable emergency equipment should be provided for both major and more routine emergencies.

2. All ports should be equipped with first-aid boxes, complete with appropriate items, stretchers and arrangements for calling an ambulance (see section 9.2.2). The contents of the boxes should be determined by an assessment, and should be checked frequently and restocked as necessary.

3. Suitable means for assisting persons who have fallen into the water should be deployed at relevant intervals along quays and other areas where portworkers may work near water (see section 3.3.6).

4. Suitable spillage control equipment should be provided in accordance with the emergency plan and be held so as to be readily available.

12. Other relevant safety matters

12.1. Environment

12.1.1. General requirements

1. Port authorities should promote sustainable development and exercise control over their activities by applying environmental protection policies to their operations.

2. This can be achieved by developing an environmental management system that will implement management and control methods that prevent or minimize damage to the environment.

3. Many of the precautions detailed elsewhere in this code of practice which are intended to prevent injury and ill health to portworkers should also be beneficial to the environment. The precautions described in this chapter are not aimed at the prevention of injury or ill health.

12.1.2. Environmental management systems

1. A suitable environmental management system should include:
— an environmental policy statement;
— suitable objectives and targets;
— allocation of environmental protection responsibilities to individual workers;
— environmental awareness training;
— environmental management practices;
— assessment of compliance and effectiveness;
— periodical review of the environmental protection programme.

2. The policy statement should clearly state the port authority's commitment to environmental protection.

3. The objectives should include:

— identifying key environmental implications for the authority's port operations in order that managers and supervisors understand the relationship between port work and the environment;

— identifying relevant key national environmental legal requirements;

— ensuring that managers and supervisors are suitably and adequately trained to carry out their environmental responsibilities;

— ensuring that assessments of possible environmental impacts are carried out before acceptance of cargoes that might have the potential to harm the environment;

— providing guidance on environmental management practices in order to minimize the risks associated with port operations;

— dealing properly with complaints.

4. International standard ISO 14001 *Environmental management systems – Specifications with guidance for use* gives general guidance on maintaining a satisfactory quality of environmental provision.

5. All workers involved in port operations should be given appropriate responsibility, within their normal functions, to ensure that the environmental management system is complied with and is successful.

12.1.3. Environmental aspects of port operations

Potential environmental concerns that may arise from port operations include:

— emissions to air;
— releases to water;
— land contamination;
— nuisance and other local community issues, e.g. noise, dust and odours;
— waste and its management.

12.1.4. Precautions

1. Every effort should be made to eliminate, restrict, control or minimize environmental concerns. Equipment design, site layout and work systems and arrangements should be organized with such objectives in mind.

2. In particular:
— every effort should be made to prevent dust or fumes becoming airborne and spreading into the atmosphere and the surrounding neighbourhood;
— every effort should be made to avoid spillage of cargoes into the water;
— any spillage on the terminal should be cleared up quickly and safely. It should not be washed into the drains where it might pollute the water or the land;
— every effort should be made to reduce noise emissions that might disturb nearby neighbourhoods, especially during work outside normal hours;
— consideration should be given to lighting arrangements that avoid undue glare disturbing nearby neighbourhoods.

12.2. Security

Port-related security issues should be addressed in accordance with the ILO/IMO code of practice *Security in ports* (2004), and, as appropriate, with the IMO's *ISPS Code, 2003 edition (International Ship and Port Facility Security Code and SOLAS Amendments, 2002)*.

References

International Labour Organization

Conventions

No. Title

111 Discrimination (Employment and Occupation) Convention, 1958

148 Working Environment (Air Pollution, Noise and Vibration) Convention, 1977

152 Occupational Safety and Health (Dock Work) Convention, 1979

155 Occupational Safety and Health Convention, 1981 [and Protocol, 2002]

161 Occupational Health Services Convention, 1985

170 Chemicals Convention, 1990

174 Prevention of Major Industrial Accidents Convention, 1993

Recommendations

No. Title

156 Working Environment (Air Pollution, Noise and Vibration) Recommendation, 1977

160 Occupational Safety and Health (Dock Work) Recommendation, 1979

164 Occupational Safety and Health Recommendation, 1981

171 Occupational Health Services Recommendation, 1985

177   Chemicals Recommendation, 1990

181   Prevention of Major Industrial Accidents Recommendation, 1993

Codes of practice or guidelines

Security in ports, ILO/IMO code of practice (2004).

Guidelines on occupational safety and health management systems, ILO-OSH 2001 (2001).

Ambient factors in the workplace, ILO code of practice (2001).

Technical and ethical guidelines for workers' health surveillance, Occupational Safety and Health Series No. 72 (1998).

Protection of workers' personal data, ILO code of practice (1997).

Accident prevention on board ship at sea and in port, ILO code of practice (second edition, 1996).

Recording and notification of occupational accidents and diseases, ILO code of practice (1996).

Safety in the use of chemicals at work, ILO code of practice (1993).

Portworker Development Programme, http://www.ilo.org/public/english/dialogue/sector/sectors/pdp/index.htm

Prevention of major industrial accidents, ILO code of practice (1991).

Protection of workers against noise and vibration in the working environment, ILO code of practice (1977).

International Organization for Standardization (ISO)

No.	*Title*
ISO 668	*Series 1 freight containers – Classification, dimensions and ratings*
ISO 830	*Freight containers – Vocabulary*
ISO 1496	*Series 1 freight containers – Specification and testing*
ISO 3874	*Series 1 freight containers – Handling and securing*
ISO 4301	*Cranes and lifting appliances*
ISO 4308	*Cranes and lifting appliances – Section of wire ropes*
ISO 4309	*Cranes – Wire ropes – Code of practice for examination and discard*
ISO 4310	*Cranes – Test code and procedures*
ISO 7752	*Lifting appliances – Controls – Layout and characteristics*
ISO 8087	*Mobile cranes – Drum and sheave sizes*
ISO 8566	*Cranes – Cabins*
ISO 9926	*Cranes – Training of drivers*
ISO 10245	*Cranes – Limiting and indicating devices*
ISO 12480	*Cranes – Safe use – Part 1: General*
ISO 14001	*Environmental management systems – Specifications with guidance for use*
ISO 14829	*Freight containers – Straddle carriers for freight container handling – Calculation of stability*
ISO 15513	*Cranes – Competency requirements for crane drivers (operators), slingers, signallers and assessors*

International Maritime Organization

Conventions

International Convention for the Safety of Life at Sea (SOLAS), 1974

International Convention for Safe Containers (CSC), 1972

Codes of practice

Code of Practice for Solid Bulk Cargoes (BC Code) (2001 edition).

Code of Safe Practice for Cargo Stowage and Securing (CSS Code) (2003 edition).

Code of Practice for Ships Carrying Timber Deck Cargoes (1991).

Code of Practice for the Safe Loading and Unloading of Bulk Carriers (BLU Code) (1998 edition).

International Maritime Dangerous Goods Code (IMDG Code) (2002 edition).

International Code for the Safe Carriage of Grain in Bulk (International Grain Code) (1991).

International Ship and Port Facility Security Code and SOLAS Amendments, 2002 (ISPS Code) (2003 edition).

Other

Recommendations on the Safe Transport of Dangerous Cargoes and Related Activities in Port Areas (1995 edition).

Recommendations on the safe use of pesticides in ships (1996 edition).

MSC/Circ. 859 Inspection programmes for cargo transport units (CTUs) carrying dangerous goods (1998).

MSC/Circ. 860 Guidelines for the approval of offshore containers handled in open sea (1998).

ICHCA International Ltd.

Container top safety, lashing and other related matters, International Safety Panel Research Paper No. 4.

Safe working on container ships, International Safety Panel Safety Briefing Pamphlet No. 8.

Other references

International Chamber of Shipping (ICS)/Oil Companies International Maritime Forum (OCIMF): *Safety guide for terminals handling ships carrying liquefied gases in bulk* (second edition, 1993).

International Association of Ports and Harbours (IAPH)/ ICS/ OCIMF: *International Safety Guide for Oil Tankers and Terminals (ISGOTT)* (fourth edition, 1996).

IAPH/ICS/OCIMF: *International safety guide for chemical tankers and terminals* (fourth edition, 1998; CD-ROM).

IMO/ILO/UN ECE Guidelines for Packing of Cargo Transport Units (CTUs) (2002 edition).

Knott, John R.: *Lashing and securing of deck cargoes* (Nautical Institute, third edition, 2002).

OECD: *Guidance concerning chemical safety in port areas* (1994).

Society of International Gas Tanker and Terminal Operations (SIGTTO): *Liquefied gas handling principles on ships and in terminals* (third edition, 2000).

UNEP/IMO: *Awareness and Preparedness for Emergencies at Local Level (APELL) for Port Areas* (1996).

Relevant web sites

ILO	www.ilo.org
IMO	www.imo.org
United Nations	www.un.org
UNEP	www.unep.org
IAPH	www.iaphworldports.org
ICHCA International Ltd.	www.ichcainternational.co.uk
ICS and ISF	www.marisec.org
ISO	www.iso.org
OCIMF	www.ocimf.com
OECD	www.oecd.org
Nautical Institute	www.nautinst.org
SIGTTO	www.sigtto.org

Appendix A

Testing of lifting appliances

A.1. General provisions

A.1.1. Every lifting appliance should be tested in accordance with the provisions of Appendix D, section D.1:

- before being taken into use for the first time;
- at least once in every period of five years if it is a lifting appliance on a ship;
- after the renewal or repair of any stress-bearing part.

A.1.2. Testing of the complete appliance is not necessary when a part is renewed or repaired, and the part is separately subjected to the same stress to which it would have been subjected if it had been tested in situ during the testing of the complete appliance.

A.1.3. Every test should be carried out:

- by a competent person;
- in daylight, provided that the latitude of the place of testing so allows; otherwise, adequate lighting should be provided;
- at a time when the wind force and/or gusting does not exceed the wind force/gusting limits for which the lifting appliance was designed;
- after all prudent precautions have been taken to ensure the safety of all persons carrying out the testing and others who may be in the vicinity at the time of the test.

A.2. Precautions before testing

A.2.1. If the stability of a ship is liable to be endangered by the test unless certain precautions, such as proper ballasting, are taken, the competent person should give notice to the master or person in charge of the ship of the date and time of the test, the amount of test load to be applied and the maximum outreach of the lifting appliance over the side of the ship. The competent person should not undertake or witness the test unless written confirmation from the master or person in charge has been received, indicating that the stability of the ship will not be endangered by the test and that the ship's deck and hatch covers are sufficiently strong to support the weight of the test load.

A.2.2. In the case of a gantry crane able to move on tracks along the deck, proper measures should be taken to ensure that movement of the crane along the track with the test load suspended can be safely controlled.

A.2.3. All temporary guys or stays for the mast or Samson posts and, where applicable, special load-slewing guys should be rigged.

A.3. Test weights

A.3.1. The weights used for making up a test load should be suitable for the purpose and be of verified weight.

A.3.2. All cast weights and, where practicable, other weights should be weighed on a machine of certified accuracy. If weighing is not practicable, the weight should be determined by calculation, the calculations retained and a copy appended to the test certificate, when issued.

A.3.3. The weight of the test load (including the weight of its lifting gear) should be not less than the figure determined from Appendix D and should not exceed it by more than 2.5 per cent.

A.4. Derricks and derrick cranes

A.4.1. All tests, except tests following the repair or renewal of a part, should be carried out by means of test weights. Tests following the repair or renewal of a part may use a spring or hydraulic weighing machine, suitably and safely anchored, provided that this may be so rigged that the part is subjected to the calculated stress to which it would be subjected if the derrick were tested by means of dead weights. When a spring or hydraulic weighing machine is used, it should be accurate to within ±2.5 per cent and the tension should be applied for a sufficiently long period to ensure that the machine's indicator remains constant for not less than five minutes.

A.4.2. A derrick should be tested with the boom at its maximum outreach corresponding to its lowest inclination to the horizontal marked upon it or to be marked upon it in accordance with section 4.3.1, paragraph 8:

– in the two extreme positions of the slewing range; and
– in the midship position.

A.4.3. In the case of a derrick, the boom and test load should be raised by the derrick's own winches with the boom in one of the positions indicated in paragraph A.4.2. It should be raised by the span winch or winches as high as possible in order that as many rope layers as possible may be reeled on the winch drum.

A.4.4. In each of the three positions indicated in paragraph A.4.2, the safe working load should be lowered at the normal lowering speed of the derrick for a distance of approximately 3 m and then braked sharply.

A.4.5. It should be demonstrated that the test load can be held stationary when the winch drive is switched off.

A.4.6. During the test it should be ascertained that, in all positions of the derrick, all parts are free to take up their appropriate positions and all ropes are running freely and are reeling up properly on the winch drums.

A.4.7. Where a derrick is designed to be used in union purchase:

– it should be tested in union purchase with its associated derrick and rigged in accordance with the ship's rigging plan. The test load should be manoeuvred throughout the working range of the union purchase and raised to such a height that the angle between the two hoist ropes is as near as possible to 120° at some position of the working range;

–   the test should be repeated with derricks rigged over the opposite side of the ship.

A.4.8. Where the derrick is fitted with a span gear winch, the winch should be tested with the derrick it serves and each sprocket should be subjected to load.

A.4.9. Upon completion of tests with the test load, each winch should be tested with its safe working load suspended and the derrick placed in various positions, such that each winch serving the derrick is subjected to loading while having the maximum working length of rope layered on its drum.

A.5. Cranes

A.5.1. Test weights only should be employed.

A.5.2. Before any test is carried out, it should be ascertained, from the manufacturer's rating or known design limitations, that the crane has been designed to withstand the impositions of the test load not only as regards its structural strength but also its stability, where this is appropriate.

A.5.3. It should be ascertained – not merely assumed – that, where appropriate, the crane is properly ballasted or counterbalanced.

A.5.4. Only an experienced competent operator should be employed during the test.

A.5.5. A mobile crane should be on level ground that is sufficiently firm to ensure that indentation or subsidence does not take place. Its outriggers (if provided) should be properly deployed and, where necessary, should be resting on timber or similar supports.

A.5.6. Tracks and rails should be checked for soundness.

A.5.7. Tyre pressures (where applicable) should be correct.

A.5.8. The safe working load limiter (section 4.1.6, paragraph 4) should be disconnected if it is of a type liable to be damaged by the test loading.

A.5.9. When a crane is tested in the "free on wheels" condition, the axle springs or shock absorbers should be chocked or locked.

A.5.10. The radius at which the test load should be applied should be measured in accordance with section 4.1.4, paragraph 3.

A.5.11. In every case, the test load should be raised sufficiently to subject every tooth in the gear wheels to loading.

A.5.12. A test load should not be deposited upon soft muddy ground, as the momentary extra resistance caused by suction between the load and the ground may be a source of danger when the load is lifted again.

A.5.13. Where a crane is secured to the structure of a building, the test should not be carried out until the owner of the building has confirmed in writing that the structure is sufficiently strong to withstand the extra strain imposed on it during the test.

A.5.14. When gantry cranes, transporters and similar lifting appliances are tested, the crane should be positioned approximately midway between any two adjacent gantry track supports. The test load should be lifted just clear of the ground and slowly traversed from one end of the bridge span to the other. In the case of a transporter, the crab or trolley supporting the test load should be slowly traversed along the entire length of the track. In the case of a gantry crane on board ship, the test load should be slowly traversed along the entire length of the track with the test load as far as possible on one side, and then again with the test load as far as possible on the other side.

A.5.15. When a mobile crane is tested, no overload test should be carried out before it has been ascertained that the crane has a sufficient margin of stability. A stability test on the crane should have been carried out by the manufacturer or, in the case of a series-produced crane, on the prototype model of that crane.

A.5.16. When any other crane is tested, such as a derrick crane that has a rigid back-stay anchored to the ground or is ballasted, an anchoring or ballast test should be carried out if the safety of the anchoring or ballasting is in doubt. The amount of test load and the manner of its application should be indicated by the manufacturer or determined by a competent person. The load should be applied with the jib or boom in a position where:

– the maximum pull on the anchorage or ballasting is achieved; or

– a reduced load at an increased radius gives an equivalent pull.

A.5.17. When a crane has a jib or boom of variable length, the test indicated in paragraph A.5.16 should be conducted with the jib or boom at its maximum length, at its minimum length, and at a length approximately midway between the maximum and minimum lengths.

A.5.18. When a boom is fitted with a fly jib or provided with more than one fly jib of different lengths, the test should be conducted on the shortest jib in combination with the main boom length that gives the greatest rated load on the fly jib. The test should also be conducted on the longest fly jib in combination with the main boom length that gives the greatest rated load on the fly jib. A further test should be conducted on the longest possible combination of main and fly jibs. Before these tests are carried out, the manufacturer's table of safe working loads for all combinations of boom length and fly jib or jibs should be made available to the competent person conducting the tests. The tests should be carried out at the position of least stability as defined by the manufacturer.

A.5.19. Where the safe working load of a crane varies according to whether it is used with stabilizing spreaders or "free on wheels", the above tests, as appropriate, should be carried out for each condition.

A.5.20. After the load test, the crane should be put through all its motions at their maximum rated speeds with the safe working load suspended, except that if the crane can freely slew through 360°, slewing should be restricted to not more than two complete turns from start to stop. All brakes should be tested.

A.5.21. Tests should also be carried out with the jib or boom at such an angle of rotation and elevation as to create the conditions of least stability as defined by the manufacturer or a competent person.

A.5.22. After the overload test, the automatic safe load indicator should be reconnected and tested by progressively applying a load to the crane until the visual and audible warnings operate. The load should be lowered to the ground on each occasion that an increment of load is applied, and then hoisted. If this is not done, the hysteresis effect in the crane's structure may result in unreliable readings.

A.5.23. All limit switches should be tested to ascertain that they are functioning correctly.

A.5.24. Upon completion of the test, the lifting appliance should be thoroughly examined by a competent person in accordance with Appendix C.

Appendix B

Testing of loose gear

B.1. General provisions

B.1.1. Every item of loose gear, other than a cargo block, should be tested in accordance with the provisions of Appendix D, section D.3.

B.1.2. Every cargo block should be tested in accordance with the provisions of Appendix D, section D.2.

B.1.3. Every item of loose gear, including cargo blocks, lifting beams and lifting frames, should be tested:

– before it is taken into service for the first time;

– after any renewal or repair of a stress-bearing part.

B.2. Physical testing equipment

B.2.1. Recording test equipment used in carrying out overload proof tests, either of assembled units or of loose gear components, should have been tested for accuracy by a competent person at least once during the 12 months preceding the test.

B.2.2. Tests should be performed with equipment that meets the standards set by the national authority, or any appropriate standard that has been verified as meeting the requirements of a national authority or other standard.

B.2.3. Machine errors should be taken into account in conducting tests.

B.2.4. A copy of relevant test reports of the testing machine should be displayed.

B.2.5. The characteristics and capacity of the recording test equipment used should be suitable for the proof tests performed.

B.2.6. Where the safe working load of the loose gear is so high or is of such a size that a testing machine is not available for applying the proof test load, or where it is not practicable to do so, the test may be carried out by suspending the gear from a suitable structure or lifting appliance and applying test weights. The weights should comply with the requirements of Appendix A, section A.3.

B.2.7. The test load on a suspended beam or frame should be applied in such a way that it will impose the maximum stress in the beam or frame. All fittings such as hooks, rings and chains should be tested independently before being fitted to the beam.

B.2.8. A pulley block should, whenever possible, be tested with its sheaves reeved, the end of the rope being properly anchored to the becket of the block. Where this is not practicable, the becket should be tested independently.

B.2.9. Slings with crate clamps, barrel hooks, plate clamps or other similar devices should be tested as nearly as possible in the manner in which they are used, i.e. at the angle at which the clamp or other device is designed to be used. The clamp or other device should be applied to a baulk of timber or special steel jig such that its holding or gripping strength is tested.

B.2.10. Upon completion of the test, the loose gear should be thoroughly examined by a competent person in accordance with Appendix C.

Appendix C

Thorough examination of lifting appliances and loose gear

C.1. General provisions

C.1.1. Where the competent person considers it necessary, parts of the lifting appliance or loose gear should be dismantled by a skilled person to the extent required by the competent person.

C.1.2. In the case of ship's gear, the examination should include associated ship's fittings such as deck eyes, mast bands, temporary stays and cleats.

C.1.3. Where the competent person considers it necessary, any parts of a lifting appliance or loose gear that may be dismantled reasonably readily should be so dismantled.

C.1.4. No lifting appliance should be used unless it has been thoroughly examined:

– after every test carried out in accordance with Appendix A, paragraph A.1.1;
– at least once during the preceding 12 months.

C.1.5. No loose gear should be used unless it has been thoroughly examined:

– after every test carried out in accordance with Appendix B, section B.1;
– after placing in service, at least once during the preceding 12 months.

C.1.6. Every part of a lifting appliance or gear specified by the competent person should be properly cleaned and prepared before the examination.

Appendix D

Test loading

D.1. Lifting appliances

The test load applied to a lifting appliance should be as follows:

Safe working load (SWL) of the appliance (tonnes)	Test load (tonnes)
Up to 20	25 per cent greater than SWL
21–50	5 tonnes greater than SWL
51 and above	10 per cent greater than SWL

D.2. Cargo or pulley blocks

The test load applied to a cargo or pulley block should be as follows:

SWL (tonnes)	Test load (tonnes)
Single-sheave block:	
All safe working loads	4 x SWL
Multi-sheave block:	
Up to 25	2 x SWL
26–160	(0.933 x SWL) + 27
161 and above	1.1 x SWL

D.3. Loose gear

The test load applied to an item of loose gear should be as follows:

SWL (of the loose gear) (tonnes)	Test load (tonnes)
Chain, hook, shackle, ring, link, clamp and similar gear:	
Up to 25	2 x SWL
26 and above	(1.22 x SWL) + 20

SWL (of the loose gear) (tonnes)	Test load (tonnes)
Lifting beam, lifting frame and similar gear:	
Up to 10	2 x SWL
11–160	(1.04 x SWL) + 9.6
161 and above	1.1 x SWL

Appendix E

Factor of safety (coefficient of utilization)

E.1. Wire rope [1]

E.1.1. For a wire rope used in the construction of a sling or forming part of a derrick, derrick crane or crane on board ship, the factor of safety should be:

–   in the case of a wire rope forming part of a sling:

Safe working load of the sling (SWL) [2]	Factor
Up to and including 10 tonnes	5
Over 10 tonnes up to and including 160 tonnes	$\dfrac{10{,}000}{(8.85 \times \text{SWL}) + 1{,}910}$
Over 160 tonnes	3

–   in the case of a wire rope forming an integral part of a derrick, etc.:

Safe working load of the derrick, derrick crane, etc. (SWL)	Factor
Up to and including 160 tonnes	$\dfrac{10{,}000}{(8.85 \times \text{SWL}) + 1{,}910}$
Over 160 tonnes	3

E.1.2. For a wire rope forming part of a crane other than on board ship, the factor of safety should be:

---

[1] See Chapter 4, section 4.4.3, in particular para. 2.

[2] In the case of multi-legged slings, this is the safe working load of the complete sling.

– calculated according to the first formula in paragraph E.1.1;
– as given in a recognized national or international standard to which the appliance has been designed and constructed.

E.2. Fibre rope

E.2.1. For cable or hawser laid synthetic or natural fibre ropes and slings, the factor of safety should be a minimum of 6. For most cargo-handling uses it is recommended that 8 be used.

E.3. Woven webbing slings

E.3.1. For synthetic woven webbing slings, the factor of safety should be a minimum of 5.

E.4. General requirements

E.4.1. The factors of safety in E.1, E.2 and E.3 should be adopted unless other requirements are laid down in a recognized national or international standard.

Appendix F

Steel quality grade mark

F.1. The quality grade mark to be placed on any steel component of loose gear in accordance with section 4.2.6, paragraph 6, should be as follows:

Quality grade mark	Grade of steel	Mean stress (N/mm^2)
L	Mild	300
M	Higher tensile	400
P	Alloy	500
S	Alloy	630
T	Alloy	800

F.2. The third column of the table above relates to the mean stress in a piece of chain made up in accordance with the appropriate ISO standard for the material at the breaking load specified in the standard.

Appendix G

Heat treatment of wrought iron

G.1. General provisions

G.1.1. Heat treatment of wrought-iron gear should consist of heating the gear uniformly in a suitably constructed muffle furnace until the whole of the metal has attained a temperature between 600°C (1,100°F) and 650°C (1,200°F), then withdrawing the gear from the furnace and allowing it to cool uniformly.

G.1.2. If the past history of wrought-iron gear is not known, or if it is suspected that the gear has been heat-treated at an incorrect temperature, before putting it to work it should be given normalizing treatment (950°-1,000°C or 1,750°-1,830°F) followed by uniform cooling. Precautions should be taken during the heat treatment to prevent excessive scaling.

G.1.3. Sling assemblies should be made of materials having similar properties.

G.1.4. However, if the assembly has some components of wrought iron and others of mild steel (e.g. mild-steel hooks permanently connected to wrought-iron chains), it should be normalized at a temperature between 920° and 950°C (1,700° and 1,750°F), removed from the furnace and cooled uniformly.

Appendix H

Marking of single-sheave blocks

H.1. General provisions

H.1.1. The safe working load (SWL) of a single-sheave block in a derrick rig should be marked in accordance with the following indications. For the sake of simplicity, the effect of friction and rope stiffness (i.e. the effort required to bend the rope round the sheave) has been ignored. In practice, the assessment of the SWL of the block, as in paragraph H.2.2, ignores friction and rope stiffness. These factors should nevertheless be taken into account when determining the resultant force on the head fittings of the heel blocks, span gear blocks and other equipment. This is the responsibility of the competent person who prepares the ship's rigging plan.

H.2. Method

H.2.1. A single-sheave block may be rigged at various positions in the derrick rig, for example in the span gear, upper and lower cargo blocks or heel blocks, and may be used with or without a becket.

H.2.2. The SWL of a single-sheave block is always assessed in accordance with one fundamental condition of loading, i.e. where the block is suspended by its head fitting and the dead weight or cargo load is attached to a wire rope passing round the sheave in such a way that the hauling part is parallel to the part to which the load is attached (figure H1). *The SWL marked on the block is the dead weight (M tonnes) that can be safely lifted by the block when rigged in this way.*

H.2.3. When the block is rigged as in paragraph H.2.2, the resultant force on the head fitting is twice the SWL marked on the block, i.e. 2M tonnes. The block manufacturer should design

476

the block in such a way that the head fitting, axle pin and strop are capable of safely withstanding the resultant force of 2*M* tonnes. Consequently, a proof load of twice the designed SWL, i.e. 4*M* tonnes, should be applied to the block.

H.2.4. When the block is rigged as a lower cargo block, i.e. when the dead weight or cargo load is secured directly to the head fitting (the block therefore being upside down) instead of to the rope passing round the sheave (figure H2), the SWL marked on the block is unchanged. The resultant force or load now acting on the head fitting is only *M* tonnes. However, as the block has been designed to withstand safely a resultant force on the head fitting of 2*M* tonnes, it follows that the block is safe to lift a dead weight or cargo load of 2*M* tonnes which gives the same stress in the block as when it is rigged as in figure H1. However, national regulations normally prohibit the use of a lifting device to lift a load in excess of the SWL marked on it, and in all but this particular case this is the correct procedure. In this particular case, and only in this case, the regulations should allow that a single-sheave block may lift twice the SWL marked on it when rigged as in figure H2 only.

H.2.5. When a suitable size of single-sheave block is to be selected for use elsewhere in the rig (for example, in a masthead span block or a derrick heel block), the maximum resultant force on the head fitting arising from the tension in the span rope should first be determined (figure H3). This force can be obtained from the rigging plan (see section 4.3.1, paragraph 1). The value of this resultant force varies according to the angle of the derrick boom to the horizontal, so that the rigging plan should show the maximum value. If this resultant force is represented by *R* tonnes, the correct block to be used at this position would be marked with an SWL equal to one-half of the resultant force (i.e. *R*/2 tonnes). However, it is extremely important to note that the shackle and link used to attach this block to the mast eye *should have and be marked with an SWL equal to* R *tonnes*. This applies to all shackles and links used for connecting blocks elsewhere in the derrick rig.

Figure H (1, 2, 3, 4, 5, 6).   Safe working loads of single-sheave blocks

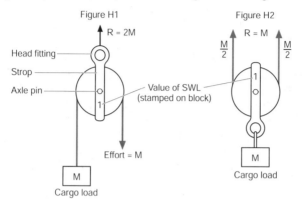

Figure H1

Head fitting

Strop

Axle pin

R = 2M

Value of SWL
(stamped on block)

Effort = M

M

Cargo load

Figure H2

R = M

$\frac{M}{2}$ $\frac{M}{2}$

M

Cargo load

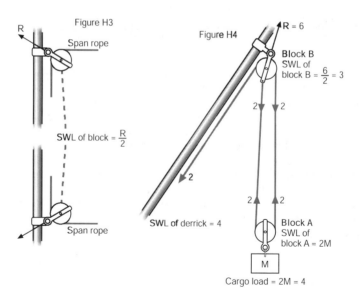

Figure H3

R

Span rope

SWL of block = $\frac{R}{2}$

Span rope

Figure H4

R = 6

Block B
SWL of
block B = $\frac{6}{2}$ = 3

2   2

2

SWL of derrick = 4

2   2

Block A
SWL of
block A = 2M

M

Cargo load = 2M = 4

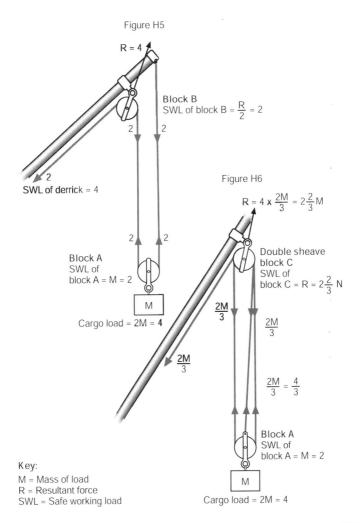

Figure H5

R = 4

Block B
SWL of block B = $\frac{R}{2}$ = 2

2   2

2
SWL of derrick = 4

2   2

Block A
SWL of
block A = M = 2

M

Cargo load = 2M = **4**

Figure H6

R = 4 x $\frac{2M}{3}$ = $2\frac{2}{3}$ M

Double sheave
block C
SWL of
block C = R = $2\frac{2}{3}$ N

$\frac{2M}{3}$ $\frac{2M}{3}$

$\frac{2M}{3}$

$\frac{2M}{3}$ = $\frac{4}{3}$

Block A
SWL of
block A = M = 2

M

Cargo load = 2M = 4

Key:
M = Mass of load
R = Resultant force
SWL = Safe working load

H.2.6. In the case of the rig shown in figure H4 (sometimes termed the "gun-tackle rig"), actual figures will serve best to explain the principle of application. Suppose the derrick is marked "SWL 4 tonnes", which is the dead weight or cargo load that can be safely handled by the derrick as a whole. It follows from paragraph H.2.4 that the lower cargo block A will be marked with a SWL of 2 tonnes, but is permitted to support a cargo load of 4 tonnes. The upper block B will have a resultant force on its head fitting of 6 tonnes (however, see paragraph H.1.1), so that the SWL of the block selected for fitting here would be $R/2$ (i.e. 6/2 or 3 tonnes). For the purpose of these examples, the fact that all the wires are not parallel has been ignored, although in practice this would not be so and the true resultant would be shown on the rigging plan.

H.2.7. Another common single-sheave block rig is shown in figure H5. The lower cargo block A would, as before, have a SWL of 2 tonnes marked on it, since this is another case where the load is directly attached to the head of the block, thus making it subject to the dispensation allowed under paragraph H.2.4, i.e. a cargo load of 4 tonnes could be lifted. The block in position B would, as explained in paragraph H.2.5, be one having a SWL of one-half the resultant force R marked on it.

H.2.8. The rig shown in figure H6 incorporates a single-sheave block (A) fitted with a becket. The upper block will in this case be a multi-sheave block and should therefore be dealt with in accordance with section 4.4.5, paragraph 12. The cargo load is attached directly to the lower block and the dispensation allowed under paragraph H.2.4 applies to it, i.e. it is stamped M tonnes but can lift $2M$ tonnes. The only effect of the becket as far as the lower single-sheave block is concerned is to reduce the tension in the wire rope from M to $2M/3$ tonnes (i.e. from 2 to $1^1/_3$ tonnes).If this were a permanent rig, a smaller size of rope would clearly be used.

H.2.9. The SWL of a single-sheave block fitted with a becket is assessed in the same way as other single-sheave blocks, i.e. according to paragraph H.2.2.

Index

Arabic numerals refer to paragraphs, not pages. Roman numerals refer to pages in the preface. Bold numerals and letters refer to diagrams.